PASTEU
QUADRANT

PASTEUR'S QUADRANT

Basic Science and Technological Innovation

Donald E. Stokes

Brookings Institution Press
Washington, D.C.

Copyright © 1997 by
THE BROOKINGS INSTITUTION
1775 Massachusetts Avenue, N.W., Washington, D.C. 20036

Library of Congress Cataloging-in-Publication data:

Stokes, Donald E.
  Pasteur's quadrant : basic science and technological innovation /
Donald E. Stokes.
    p.   cm.
  Includes bibliographical references and index.
  ISBN 0-8157-8178-4 (cloth).—ISBN 0-8157-8177-6 (pbk.)
    1. Science and state—United States.   2.Technology and state-
United States.   3. Science—History.   4. Research—History.
5. Technological innovations—History.   6. Paradigm (Theory of
knowledge)   I. Title.
Q127.U6S757   1997
338.97306   dc21   97-4807
                                                                   CIP

9 8 7 6 5 4 3 2 1

The paper used in this publication meets the minimum requirements of the
American National Standard for Information Sciences—Permanence of Paper
for Printed Library Materials, ANSI Z39.48-1984.

Typeset in Sabon

Composition by Harlowe Typography, Inc., Cottage City, Maryland

Printed by R. R. Donnelley and Sons, Co., Harrisonburg, Virginia

FOREWORD

MORE THAN FIFTY YEARS ago, Vannevar
Bush released his enormously influential re-
port, *Science, the Endless Frontier*, which asserted a dichotomy
between basic and applied science. This view was at the core of
the compact between government and science that led to the
golden age of scientific research after World War II—a compact
that is currently under severe stress. In this book, Donald E. Stokes
challenges Bush's view and maintains that we can only rebuild the
relationship between government and the scientific community
when we understand what is wrong with that view.

Stokes begins with an analysis of the goals of understanding
and use in scientific research. He recasts the widely accepted view
of the tension between understanding and use, citing as a model
case the fundamental yet use-inspired studies by which Louis Pasteur laid the foundations of microbiology a century ago. Pasteur
worked in the era of the "second industrial revolution," when the
relationship between basic science and technological change as-
sumed its modern form. During subsequent decades, technology
has been increasingly science based—with the choice of problems
and the conduct of research often inspired by societal needs.

On this revised, interactive view of science and technology, Stokes builds a convincing case that by recognizing the importance of use-inspired basic research we can frame a new compact between science and government. His conclusions have major implications for both the scientific and policy communities and will be of great interest to those in the broader public who are troubled by the current role of basic science in American democracy.

Having put the final touches on his manuscript, Donald E. Stokes died of acute leukemia on January 26, 1997. At the time of his death, he was professor of politics and public affairs in the Woodrow Wilson School of Public and International Affairs at Princeton University. Stokes served as dean of the school between 1974 and 1992.

At Brookings, Theresa Walker edited the manuscript, Matthew Atlas and Tara Adams Ragone verified it, Inge Lockwood proofread it, and Julia Petrakis prepared the index.

The views expressed in this book are solely those of the author and should not be ascribed to the trustees, officers, or other staff members of the Brookings Institution.

Michael H. Armacost
President

July 1997
Washington, D.C.

PREFACE

THE PROBLEM I EXPLORE in this book first caught my eye when I was dean of the graduate school at the University of Michigan, a role that was, among other things, a walking subscription to *Scientific American*.

As I made my rounds of a number of scientific fields I was struck by how often a gifted scientist would talk about the goals of research—especially the relationship between the quest of fundamental understanding on the one hand and considerations of use on the other—in a way that seemed to me odd. Odd and unhelpful, since my preceptors' view of this relationship and of the relationship between the categories of basic and applied research derived from these goals kept them from seeing things I felt they needed to see.

This reaction was strongly reinforced when I served for several years on a council advising the director of the National Science Foundation and heard this same formulation on a number of occasions. One morning, as an eminent scientist again voiced these beliefs, I so startled the council with an alternative view that my ideas were projected as an overhead slide at the beginning of the afternoon session. An updated version of this slide appears in some of the figures in chapter 3. The Foundation widened its complicity by publishing a statement of the argument I sent the director.[1] I had a chance to explore other parts of the problem when I chaired a National Research Council panel that studied the federal government's support of research on social problems.[2]

My interest in the problem was kept alive by serving for a number of years as dean of the Woodrow Wilson School at Princeton. The research efforts of this school so clearly involved the interplay of understanding and use in the social sciences that no one could lead such a unit without thinking very deeply about this relationship, and I draw liberally here on the experience of the school's Office of Population Research and Research Program in Development Studies. Eventually I came to believe that these issues deserved to be explored in a book-length work.

It took somewhat longer to be convinced that I should write it, since the early chapters deal with elements of the history of science and of intellectual history in which I began with no particular advantage. But the issues I raise have implications for three things in which I have been directly involved—the building of research agendas, the creation of institutional settings for research, and the channeling of research support. The latter chapters of this book trace the implications of a revised view of the relationship between basic science and technological innovation for each of these areas of science policy.

No one should write such a book, least of all a book that cuts across a number of fields, without a clear idea of who might read it. The argument I set out is of natural interest to those who deal with science and technology policy in and outside of government and to members of the scientific community within the universities, the government, and the free-standing research institutes and firms. Since I draw on the experience of several of the countries of the industrial world, my argument may be of interest to the science and policy communities in these countries as well. And since I pass familiar light through new prisms, my argument may also be of

1. Donald E. Stokes, "Making Sense of the Basic/Applied Distinction: Lessons for Public Policy Programs," in *Categories of Scientific Research*, papers presented at 1979 National Science Foundation seminar, Washington.

2. The principal report of this study is National Research Council, *The Federal Investment in Knowledge of Social Problems* (National Academy of Sciences, 1978).

interest to historians of science and historians of ideas, however synthetic my scholarship in these fields may be.

Social scientists will recognize this as a work of social science. Indeed, my political science colleagues will have no difficulty seeing it as a work of political and institutional analysis. But my argument extends to research in all scientific fields—including the physical sciences and engineering, the biological and biomedical sciences, and the social sciences—since there is a unity to science in the respects that are critical to my argument. But this carries no implication that the sciences are in all respects the same; and certainly none that social science is as close to natural science as biology, say, is to physics.

I could not have sharpened my argument without the help of many friends and colleagues. Too many have lent their wisdom and encouragement for me to acknowledge them all. My special appreciation goes to a number of my Princeton colleagues, including Clinton Andrews, Peter Eisenberger, Harold Feiveson, Charles Gillispie, Frank von Hippel, Daniel Kammen, Walter Kauzmann, Michael Mahoney, Harold Shapiro, Robert Socolow, Thomas Spiro, Thomas Stix, and Norton Wise; if nothing else, this book is a tribute to the intellectual commerce within this university. Of the many members of the "invisible college" who have offered insight and encouragement from a distance, a special debt is owed to the remarkable Harvey Brooks, who read the manuscript with care and deep insight. I would also especially mention Max Kaase, Richard Nelson, Stephen Nelson, Albert Teich, and John Servos. I have benefited from the help of a number of people in government, including Jennifer Sue Bond, Patricia Garfinkel, and Carlos Krytbosch.

Carolyn North, my research assistant for two periods, gathered the materials for this work with intelligence, insight, and care. Mary Huber prepared the ground for this effort, and Betsy Shalley-Jensen, Robert Sprinkle, Frank Hoke, Chris Thompson, Michael McGovern, and Esra Diker skillfully grasped the baton as it was passed to them. I am greatly indebted to each of them.

I want to acknowledge my debts to four research organizations that have lent me invaluable assistance. In the autumn and winter of 1992–93 the Research Institute of International Trade and Industry in Tokyo helped to open a window to Japan's experience with science and technology policy. In the spring of 1993 the Royal Society of London and the Science Policy Research Unit of the University of Sussex deepened my insight into the experience of Britain and Europe. I am very grateful indeed to Peter Collins and Mike Ringe at the Royal Society and to Christopher Freeman, Michael Gibbons, Diana Hicks, Ben Martin, Keith Pavitt, Margaret Sharp, and their colleagues at Sussex.

Finally, Bruce MacLaury, the president of the Brookings Institution, and Thomas Mann, the director of its Governmental Studies Program, offered unfailing support as I pursued a project that has ranged freely over the fields of science and technology, the several millennia of the Western experience of science and scientific philosophy, and the contemporary approaches to science and technology policy taken by the major countries of the industrial world. I am grateful to them and to Paul Peterson, Thomas Mann's predecessor, and my other interim Brookings colleagues. Because the Brookings Institution's own mission so clearly involves the goals both of understanding and use, it proved to be an ideal location in which to reduce my analysis to a written text.

Donald E. Stokes
September 1996

CONTENTS

PASTEUR'S
QUADRANT

1 | STATING THE PROBLEM

THE FORCES UNLEASHED by the scientific revolution of the seventeenth century and the industrial revolution of the nineteenth century helped create the modern world. But as the twentieth century draws to a close, the measures adopted by the leading industrial countries to harness these twin engines of modernization are in considerable disarray.

A half century earlier, the major scientific countries, led by the United States, emerged from World War II with policies that were based on a widely accepted view of the role of basic science in technological innovation, and these policies remained remarkably stable over several decades. But this postwar framework has come under intense pressure in recent years, and searching reviews of science and technology policy have been undertaken in the United States and the other industrial countries, including Britain, France, Germany, and Japan.

The reason for this change most often cited in the United States is the end of the cold war. Although this reason is particular to America, it was inevitable that the release of the billions of research and development dollars impounded by the Soviet confrontation would raise questions about the federal investment in science and technology. The compact between science and government reached in the early years of the cold war has come unstuck, and the scientific and policy communities are actively canvassing for the terms of a fresh agreement.

Yet it would be a mistake, even in America, to attribute the present disarray simply to the vanishing Soviet threat. At a deeper level the postwar bargain has been undermined by weaknesses in the postwar beliefs about the relationship between science and technology. Well before the end of the cold war these limitations had inspired considerable skepticism about the prevailing policies, which by no means had clear sailing through the decades of the Soviet threat.[1] We need a more realistic view of the relationship between basic science and technological innovation to frame science and technology policies for a new century. Before the argument of these chapters fully unfolds, the search for a new understanding will raise the deepest issues surrounding the role of basic science in a political democracy.

Forging the Postwar Paradigm

Late in 1944, a year before the end of World War II, Franklin D. Roosevelt asked Vannevar Bush, his director of the wartime Office of Scientific Research and Development, to look ahead to the role of science in peacetime. Before Bush could file his report, Roosevelt was dead and the country was readying the grim capstone it would place on its scientific success in the war by exploding an atomic device in the New Mexico desert. But Bush's report, *Science, the Endless Frontier,* did what Roosevelt had asked and set out a vision of how the nation could sustain its investment in scientific research when the war was over. Half a decade later, the view of basic science and its relation to technological innovation set out in the Bush report became a foundation of the nation's science policy for the postwar decades.[2]

The reasons for the profound influence of the report lay less in Bush's detailed policy blueprint than his framework for thinking about science and technology as he and his colleagues sought to extend the government's support of basic science into peacetime while drastically reducing the government's control of the performance of research. Indeed, this conceptual framework came to have a greater significance than Bush had intended as his plan for a National Research Foundation as broad as the wartime Office of Scientific Research and Development foundered in the postwar years and the scientific community fell back on Bush's conceptual premises to achieve its goals.

In a style reminiscent of Francis Bacon, Bush compressed these premises into two aphorisms. Each was cast in the form of a statement about basic research, a term he coined. The first was that "basic research is performed without thought of practical ends." Although this sounds like a definition and is often been taken to be one,[3] Bush went on to make clear that the defining characteristic of basic research is its contribution to "general knowledge and an understanding of nature and its laws."[4] His first canon about basic research instead expressed the belief that the creativity of basic science will be lost if it is constrained by premature thought of practical use. Bush saw an inherent tension between understanding and use as goals of research and, by extension, an inherent separation between the categories of basic and applied research that are derived from these goals. Indeed, he went on to endorse a kind of Gresham's Law for research, under which *"applied research invariably drives out pure"* if the two are mixed.[5] This tension is nicely captured by the familiar idea of a spectrum between basic and applied research, the one-dimensional graphic that came to represent the static version of the postwar paradigm; research cannot be closer to one of the poles of this continuum without being farther away from the other.

If Bush's first aphorism laid the foundation for the static version of the postwar paradigm, his second laid the foundation for the dynamic version. "Basic research," he wrote, "is the pacemaker of technological progress."[6] He expressed in this the belief that if basic research is appropriately insulated from short-circuiting by premature considerations of use, it will prove to be a remote but powerful dynamo of technological progress as applied research and development convert the discoveries of basic science into technological innovations to meet the full range of society's economic, defense, health, and other needs. The equally one-dimensional image that came to represent this dynamic version of the postwar vision is the familiar "linear model," with basic research leading to applied research and development and on to production or operations, according to whether the innovation is of a product or process.

Bush's view of the relationship between fundamental science and technological innovation contained an additional element, closely related to his second canon of basic research—that those who invest in basic science will capture its return in technology as the advances of science are converted into technological innovation by the processes of technology transfer. He asserted this belief in an obverse form, saying that *"a nation which depends upon others for its new basic scientific knowledge will be slow in its industrial progress and weak in its competitive position in world trade."*[7]

From a distance of five decades, we can only admire Bush's achievement. He chose for his two canons ideas that have, as we will see, a deep resonance in the Western tradition of science and scientific philosophy, one of them dating back to the invention of scientific inquiry in classical antiquity, the other to the beliefs about science voiced by Francis Bacon and others in early modern Europe. Bush wove these ideas into a plan for promoting the country's goals while allowing its scientists to pursue basic research wherever it might lead. As this plan was absorbed into policy in the postwar decades, it did allow the country to advance its goals as many of its ablest and most highly trained scientists pursued basic research wherever it led, at public expense.

We can also only admire how well Bush chose his moment for converting the kinetic energy of science's wartime success into the potential energy of the government's standing commitment to science in peacetime. The world stood awestruck before the power of science to bring the Pacific war to its astonishing close. Exploding the atomic bomb created a remarkable opening in the national consciousness for a report that charted the future role of science in the nation's life. As a result, Bush's canons left a deep impression and provided the dominant paradigm for understanding science and its relation to technology in the latter part of the twentieth century. These ideas can still be heard in the scientific and policy communities, the communications media, and the informed public. And America's leadership in postwar science has given them wide circulation in the international community.

But the influence of this paradigm has come at a price, since it obscures as well as reveals. Bush's canon on the essential goal of basic research gives too narrow an account of the motives that inspire such work. And his canon on the importance of basic research for advances in technology also gives too narrow an account of the actual sources of technological innovation. As a result, this paradigm has made it more difficult to think through a series of policy issues that require a clear vision of the goals of scientific research and of the relationship of scientific discovery to technological improvement.

These limitations are more troublesome today than they were in the postwar world, when Bush's outlook seemed to be validated by America's preeminence in science and in technology. A number of countries, including the United States, see their investment in science as a means of remaining competitive in the global economy. This shift poses new questions about whether a nation can gain a competitive edge by capturing the fruits of its basic research in new technology, or whether these fruits become part of a common fund of scientific knowledge that can be exploited by its rivals as well. Indeed, the changing context of science and technology policy has put intense pressure on the idea of basic research as a remote dynamo of progress. Although a general critique of the prevailing paradigm has yet to appear, today's circumstances make it increasingly timely.

A fresh look at the goals of science and their relation to technology is what this book is about. It reexamines the link between the drive toward fundamental understanding and the drive toward applied use, shows how this relationship is often misconceived and the price we pay for this, proposes a revised view of the interplay of these goals of science and of the relationship between basic science and technological innovation, and shows how this revision could lead to a clearer view of several aspects of science and technology policy. This first chapter describes the problematic elements of the postwar paradigm. Chapter 2 examines the history of ideas to resolve the paradox posed by the widespread acceptance of this paradigm. Chapter 3, the pivot of the argument, sets out a revised view of the relationship between understanding and use as goals of research—and between the categories of basic and applied research derived from these goals—offering a quite different view of the links between basic science and technological innovation. Chapter 4 shows how this revised view could help renew the compact between science and government. Chapter 5 seeks a process by which American democracy can build agendas of use-inspired basic research by bringing together judgments of research promise and societal need.

The analysis begins with the nature of basic and applied research, since the relationship of research inspired by the quest for understanding and research inspired by considerations of use helps to define our essential problem. As the analysis unfolds, we will see where the prevailing paradigm is faithful to, and where it distorts, the real interplay of the goals of science and the links between basic science and technological innovation.

The Concepts of Basic and Applied Research

Research proceeds by making choices. Although the activities by which scientific research develops new information or knowledge are exceedingly varied, they always entail a sequence of decisions or choices. Some of these have to do with the choice of problem area or particular line of inquiry, some with the construction of theories or models, some with the derivation of predictions, deductions, or hypotheses, some with the development of instruments or measures, some with the design of experiments and the observation of data, some with the use of analytic techniques, some with the selection of follow-on inquiries, some with the communication of the results to other scientists. Harvey Brooks caught this universal aspect of research when he said that "any research process can be thought of as a sequential, branched decisionmaking process. At each successive branch there are many different alternatives for the next step."[8] The distinction between basic and applied research turns on the criteria that govern the choice among these alternatives.

Three observations set our argument in motion. The initial observation is this:

The differing goals of basic and applied research make these types of research conceptually distinct.

On any reasonable view of the goals of basic and applied research, one cannot doubt that these categories of research are conceptually different. The defining quality of basic research is that it seeks to widen the understanding of the phenomena of a scientific field. Although basic research has been defined in many ways and involves the extraordinarily varied steps just suggested, its essential, defining property is the contribution it seeks to make to the general, explanatory body of knowledge within an area of science. In keeping with this conception, the Organization for Economic Cooperation and Development defines basic research as "experimental or theoretical work undertaken primarily to acquire new knowledge of the underlying foundation of phenomena and observable facts," although the OECD definition adds a disclaimer as to practical use to which we will return.[9] Sometimes basic research is defined in terms of certain correlates on which it differs from applied research, such as originality, the freedom of researchers, peer evaluation of published results, and length of time between discovery and practical use. But these corollary properties ought not to be taken for the characterizing quality of basic research—its thrust toward a wider understanding of the phenomena of a field.

This quality can be found in any number of examples from the annals of research. One that is useful for the further discussion is supplied by the study that launched the scientific career of Louis Pasteur when the enigma of racemic acid caught his eye as a student at the *École Normale Supérieure* in Paris. The Berlin chemist Mitscherlich had found that two remarkably similar acids, tartaric and paratartaric (or racemic) acid, had very different actions on light, since tartaric acid rotated a plane of polarized light through a characteristic angle whereas racemic acid did not—despite the fact that the two appeared to be identical in chemical composition, crystalline form, specific weight, and other properties.

Mitscherlich's report of this anomaly plunged Pasteur into the search for an explanation. When he turned his microscope on crystals made from racemic acid he found that they were of two forms, one identical to crystals of tartaric acid, the other their mirror image. Separating the two he found that a solution of the crystals identical to the tartrates rotated the plane of polarized light exactly as tartaric acid did, whereas a solution of the mirror-image crystals rotated the plane by the identical angle in the *op-posite* direction. A solution with equal proportions of the two was optically neutral, deflecting the plane of polarized light not at all. Pasteur's excited *"tout est trouvé"* took its place in the litany of scientific discovery. He had indeed solved the problem by showing that racemic acid is composed of two isomeric forms whose equal and opposite actions on light canceled each other when the two were combined. His research, guided at each stage by the quest of understanding, had extended the frontiers of crystallography.

If basic research seeks to extend the area of fundamental understanding, applied research is directed toward some individual or group or societal need or use. This quality is illustrated well by an applied problem from Pasteur's subsequent career, his effort to cope with the persistent difficulties experienced by those who made alcohol from beets. These difficulties led an industrialist in the Lille region to seek his help. As the dean of the local Faculty of Science Pasteur had encouraged his students to do practical work in industry before pursuing industrial careers. He visited a factory and took samples of the fermenting beet juice to his laboratory for microscopic examination.

Threading his way through a maze of scientific misconceptions, Pasteur identified the microorganisms responsible for fermentation and showed that they could survive without free oxygen—indeed, that they produced the alcohol resulting from fermentation by wresting oxygen from the sugar molecules in the fermenting juice. This insight gave his industrial clients an efficient means of controlling fermentation and limiting spoilage. James Bryant Conant, in his case study of this work by Pasteur, notes that one of the most valuable properties of applied research is "reducing the degree of empiricism in a practical art."[10] Pasteur's study dramatically reduced the degree of empiricism in the industries using fermentation.

If the goal of basic research is, in a word, understanding, and of applied research, use, it cannot be doubted that these types of research are conceptually or analytically different. But the prevailing view of scientific research often includes a further element, one that leads to the second observation that sets our argument in motion:

An inherent tension between the goals of general understanding and applied use is thought to keep the categories of basic and applied research empirically separate.

A particular piece of research will, on this view, belong to one or the other of these categories but not both. This was Bush's view in *Science, the Endless Frontier* when he spoke of "a perverse law governing research," under which "*applied research invariably drives out pure.*"[11] An inherent conflict between the goals of basic and applied research is thought to preserve an empirical boundary between the two kinds of inquiry.

This view did not spring Athena-like from Bush's brow after the war; in chapter 2 the idea of pure inquiry is traced through two millennia. But the perceived conflict between the goals of basic and applied research has rarely been so clearly spelled out as it was in Bush's report. The separateness of basic and applied research implied by this presumed conflict is an idea that is woven into the dominant paradigm of science and technology policy and perceptions of science held in government, the research community, and the communications media.[12] It is impossible to go through the commentaries on science of recent decades without sensing how deeply this idea pervades our outlook on scientific research. The belief that basic and applied research are separate categories also has a considerable history, and chapter 2 shows how it has been reinforced by the institutional development of science in Europe and America in the nineteenth and twentieth centuries.

Static and Dynamic Forms of the Paradigm

The belief that understanding and use are conflicting goals—and that basic and applied research are separate categories—is captured by the graphic that is often used to represent the "static" form of the prevailing paradigm, the idea of a spectrum of research extending from basic to applied:

This imagery in Euclidean one-space retains the idea of an inherent tension between the goals of understanding and use, in keeping with Bush's first great aphorism, since scientific activity cannot be closer to one of these poles without being farther away from the other.

The distinctness of basic from applied research is also incorporated in the dynamic form of the postwar paradigm. Indeed, the static basic-applied spectrum associated with the first of Bush's canons is the initial segment of a dynamic figure associated with Bush's second canon, the endlessly popular "linear model," a sequence extending from basic research to new technology:

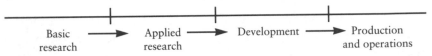

The belief that scientific advances are converted to practical use by a dynamic flow from science to technology has been a staple of research and development (R&D) managers everywhere. Bush endorsed this belief in a strong form—that basic advances are the *principal* source of technological innovation, and this was absorbed into the prevailing vision of the relationship of science to technology. Thus an early report of the National Science Foundation commented in these terms on this "technological sequence" from basic science to technology, which later came to be known as "technology transfer":

   —The technological sequence consists of basic research, applied research, and development. . . .
   —Basic research charts the course for practical application, eliminates dead ends, and enables the applied scientist and engineer to reach their goal with maximum speed, directness, and economy. Basic research, directed simply toward more complete understanding of nature and its laws, embarks upon the unknown, [enlarging] the realm of the possible.
   —Applied research concerns itself with the elaboration and application of the known. Its aim is to convert the possible into the actual, to demonstrate the feasibility of scientific or engi-

neering development, to explore alternative routes and methods for achieving practical ends.

—Development, the final stage in the technological sequence, is the systematic adaptation of research findings into useful materials, devices, systems, methods, and processes. . . .

From these definitions it is clear that each of the successive stages depends upon the preceding [one].[13]

If production and operations, the final stage of converting basic science into new products or processes, is added, the linear model is produced. This sort of dynamic linear-model thinking gave rise to the Department of Defense's categories for R&D, which soon accounted for the major share of postwar federal spending on research. Together with its equally linear static corollary, the basic-applied spectrum, this dynamic linear image provided a general paradigm for interpreting the nature of research, one that is remarkably widespread in the scientific and policy communities and in popular understanding even today.[14]

The diffusion of this paradigm in the postwar world is suggested by another voice in another place. Keith A. H. Murray, longtime rector of Lincoln College, Oxford, and chairman of Britain's University Grants Committee, instructed Australia's prime minister, Robert Menzies, and the government colleagues of Menzies on the needs of Australia's universities in the second decade after the war. The 1957 report of the Murray Committee said in part:

> It is obvious that most of the basic secrets of nature have been unravelled by men who were moved simply by intellectual curiosity, who wanted to discover new knowledge for its own sake. The application of the new knowledge usually comes later, often a good deal later; it is also usually achieved by other men, with different gifts and different interests.[15]

This declaration expresses both the belief that basic and applied research are separate ventures, pursued by different people "with different gifts and different interests," and the belief in the priority in time of the discoveries of basic science.

As the validity of these beliefs is examined, one must remember that the goals defining the categories of basic and applied research by no means exhaust the motives driving the scientific enterprise. Those who have offered general or particular accounts of the motives of research scientists paint an extraordinarily diverse portrait of the actual incentives for research. Some of these are strongly joined to the normative structure of science, as Robert K. Merton's classic study of the race for priority in scientific publication shows.[16] But the presence of other motives for research does not diminish the importance of deeply probing the relationship between the goals of understanding and use, since the postwar paradigm is characterized by the belief that these goals are necessarily in tension and the categories of basic and applied research necessarily separate as well as by the belief that innovations in technology have their source in advances in basic science.

The Experience of Science

It is possible to form a very different view of these relationships from the annals of research, and the third observation completes the statement of the problem:

> The belief that the goals of understanding and use are inherently in conflict, and that the categories of basic and applied research are necessarily separate, is itself in tension with the actual experience of science.

Although a great deal of research is wholly guided by one or the other of the goals of understanding and use, some studies of great importance show that the successive choices of research are influenced by *both* these goals.

This possibility is strikingly illustrated by the rise of microbiology in the nineteenth century; the examples from Pasteur's work were deliberately chosen. No one can doubt that Pasteur sought a fundamental understanding of the process of disease, and of the other microbiological processes he discovered, as he moved through the later studies of his remarkable career. But there is also no doubt that he sought this understanding to reach the applied goals of preventing spoilage in vinegar, beer, wine, and milk and of conquering *flacherie* in silkworms, anthrax in sheep and cattle, cholera in chickens, and rabies in animals and humans.

This mix of goals was not visible in the young Pasteur. The 22-year-old chemist who immersed himself in the enigma of racemic acid was engaged in a pure quest of understanding. Yet as Pasteur went to work on this enigma, he caught sight of a further puzzle, the question of why racemic acid mysteriously appeared in some places and not in others. He strongly suspected that microscopic agents were at work, and this conjecture greatly enhanced his interest in the microorganisms he found responsible for fermenting beet juice into alcohol in his studies at Lille. As he pursued this research, he began to fashion a framework for understanding a whole new class of natural phenomena, and he obtained the strikingly original result that certain microorganisms were capable of living without free oxygen. This work launched his assault on the medieval doctrine of the spontaneous generation of life and led to the brilliant later studies in which he developed the germ theory of disease. Hence, as Pasteur's scientific studies became progressively more fundamental, the problems he chose and the lines of inquiry he pursued became progressively more applied.

The problem of deriving alcohol from beet juice makes this point well. Pasteur's work on this problem is, as Conant noted, a distinguished example of applied research, a highly successful effort to improve the technology of fermentation. But the study that Conant called a prime example of applied research was at the same time a distinguished example of basic research. This blend characterized virtually the whole of Pasteur's later career. He probed ever more deeply into the processes of microbiology by accepting applied problems from a Lille industrialist, from the minister of agriculture, even from the Emperor Napoleon III—and, in a case that did much to build the Pasteur legend, from the distraught mother of a child bitten by a rabid dog. Many of his detailed lines of inquiry, such as the experiments by which he developed the process of the "pasteurization" of milk or his experiments in growing attenuated bacterial strains to immunize patients from disease, are unintelligible apart from his applied goals. The mature Pasteur never did a study that was not applied, as he laid out a whole new branch of science.[17]

Pasteur's example was by no means unique. Across the English Channel, Kelvin's physics was inspired by a deeply industrial view and the needs of Empire.[18] Across the Rhine, the German organic chemists were making fundamental advances to lay the basis of Germany's chemical dye industry and, later, pharmaceuticals and, by the time of Staudinger, plastics. In America, Irving Langmuir earned a Nobel Prize in 1932 for working out the physical chemistry of the surfaces of the components being manufactured by the nascent electronics industry. In the century following Pasteur every branch of science recorded advances that were partly inspired by considerations of use.

Certainly, the modern biological sciences are difficult to bring within the traditional, either-or view of basic and applied research. The revolution in molecular biology has posed questions, such as how interferon works, that were enormously important both for the advance of fundamental knowledge on recombinant DNA and for major applications—some of which will be immensely profitable. A similar observation can be made about the nonmolecular parts of modern biology. Some of the fundamental problems in population dynamics, such as the biology underlying recruitment processes and stock densities in fish, have applications that have inspired the most innovative research in the field.

Nor is this fusion of goals unique to the biological sciences. The goals of understanding and use are as closely linked in a cluster of earth sciences as they are in the life sciences. The fields of seismology, oceanic, and atmospheric science were brought into being partly by the traditional dread of earthquake, storm, drought, and flood, and their science has been enriched by such distinctively modern concerns as global warming and the detection of nuclear blasts.

The separation of "pure" physical science from engineering has, as chapter 2 notes, reinforced the impression of the inherent separateness of basic from applied science, and many of those who work on the physical science side of this divide see the split as validating the idea of an inherent separation of pure from applied. But a number of those who work on the engineering side of this divide see their fields, with some justice, as providing a home for research that is driven by the goals both of basic understanding and applied use.

A revealing example is the advance of physical chemistry achieved by W. K. Lewis, A. A. Noyes, and G. N. Lewis at the Massachusetts Institute of Technology (MIT) after World War I.

MIT had played an important role as the subject matter of chemical engineering was reorganized around such generic processes as distillation, filtration, and absorption. Noyes and the two Lewises extended these developments by exploring such phenomena as heat exchange and high temperature chemistry, reactions at high pressure, and gas absorption at a still more general and abstract level, creating an impressive new school of physical chemistry. But they also provided a far stronger base of knowledge to meet the needs of their industrial constituency. If the goal of general understanding powerfully guided this work, it did so without in any way weakening the goal of use. Gillispie cites this research as one of three early cases in which American science "contributed fundamentally to . . . the form and content of a discipline," noting that the significance of this research was partly

> that the summons to generality should have been heard, not in the conventional reaches of basic science, but in a realm of industrial science—heard and answered.[19]

These developments are a clear example of research in physical science driven by the joint goals of understanding and use. The Lewises, as much as Pasteur, were attuned to both.

One of the clearest moments when modern physical research fused the goals of understanding and use—the development of atomic weapons during the war—also serves as a revealing example of the tendency to view the experience of science in terms of the prevailing paradigm. The unease of the scientists who worked in the Manhattan Project was partly due to the moral ambiguity of atomic weapons. But their unease was also due to the tension between the need to work under tight security in a highly organized setting and their prewar vision of basic research as conducted by individual scholar-scientists who were free to share their discoveries with a far-flung company of professional peers, as well as to the tension between the wartime constraints on the choice of research problems and their belief in the autonomy of fundamental research.[20]

The result was a strong tendency after the war to remember the Manhattan Project as a gigantic exercise in applied research and development, and not as a remarkable effort in basic research as well. J. Robert Oppenheimer, who directed the Los Alamos Scientific Laboratory, declared that

> the things we learned [during the war] are not very important. The real things were learned in 1890 and 1905 and 1920, in every year leading up to the war, and we took this tree with a lot of ripe fruit on it and shook it hard and out came radar and atomic bombs. . . . The whole spirit was one of frantic and rather ruthless exploitation of the known; it was not that of the sober, modest attempt to penetrate the unknown.[21]

Henry DeWolf Smyth, who wrote the authoritative report on the development of the atomic bomb, viewed the war years as "a period of almost complete stagnation" and felt that in consequence "the fountainhead of all our future scientific developments has run dry."[22]

This commentary almost certainly says as much about the tacit assumptions held by these distinguished scientists as it does about the actual experience of science during the war. The Manhattan Project may not have achieved scientific breakthroughs as fundamental as those by Niels Bohr and others in the prewar decades. But it reached its goal only by promoting the development of the field of nuclear physics and by laying down a base of knowledge about the fundamental phenomena of the field, such as the probability of neutrons being captured by nuclei at various neutron energies and the neutron scattering and absorption of various nuclear isotopes. Nuclear physics greatly benefited from the discovery and study of many new isotopes among uranium and plutonium fission and neutron-capture products. The advances in basic science achieved by the Manhattan Project also nourished later work in related fields. The understanding it achieved of implosion phenomena, for example, contributed important insights to the later study of supernovae. It would be difficult to explain the full excitement of the project for scientists of the quality of Luis W. Alvarez, Hans A. Bethe, Enrico Fermi, John Louis von Neumann, Robert Oppenheimer, Isidor Isaac Rabi, Glenn T. Seaborg, Emilio Gino Segre, Leo Szilard, Edward Teller, Stanislaw M. Ulam, Victor Frederik Weisskopf, and Eugene Paul Wigner if its basic science had not posed an extraordinary intellectual challenge. It is more reasonable to see this wartime experience not as negating the opportunity for basic research but as channeling this research toward an overriding national goal.

The social sciences also offer striking cases of advances that were driven by the desire to extend basic knowledge and to reach applied goals. A conspicuous example is the unfolding of macroeconomic theory in the hands of John Maynard Keynes and his heirs. Keynes wanted to understand the dynamics of economies at a fundamental level. But he also wanted to abolish the grinding misery of economic depression. Although our understanding of the economy remains unfinished, and sustained growth only partially realized, we could not miss the fusion of goals in this line of social science research.

Recent research on the sources of economic development illustrates this fusion of goals in social research aimed at improved practice. Those working in the field of economic development have wanted to raise many of the peoples of the earth above the poverty line. But they have also sought to understand the sources of economic growth at a fundamental level. There is a Pasteur-like clarity to the link between understanding and use in the work of Arthur Lewis, a pioneer in the field of economic development whose contributions were recognized by a Nobel Prize. Since he was from the third world, he had an intense desire to help solve the economic problems confronting the developing countries in a postcolonial era. And the third world students crowding into his classrooms in Manchester and Princeton virtually begged for tools of economic development they could use to improve the lot of their countries when they returned home. Yet Arthur Lewis discovered his most important contribution to development economics, his two-sector model of development, only by probing the deepest intellectual puzzles in economics, as he himself described in his Nobel lecture.[23]

The rise of modern demography furnishes a further clear example of the fusion of understanding and use in the social sciences. Those who laid the foundations of this field had an outlook similar to the view of those who pioneered the field of macroeconomics. They wanted to understand the sources of population change at a fundamental level. But they also saw population change as a problem that required concerted, informed action. The case is an interesting one because this problem focus became, if anything, sharper as the quest for understanding moved to deeper levels. In demography's early years its research agendas came under heavy pressure from those who wanted to support quick action programs. At this stage a small core of research demographers pulled back and pursued a far more fundamental research agenda, partly by developing highly sophisticated mathematical models of population replacement. The worth of this strategy of pursuing applied goals through fundamental understanding was borne out when these models were refined after World War II for the limited fertility and mortality data of third world countries—and revealed for the first time, fewer decades ago than we now remember, the staggering force of the population explosion that lay in store.[24]

Science and Technology

The examples from the history of science that contradict the static form of the postwar paradigm call into question the dynamic form as well. If applied goals can directly influence fundamental research, basic science can no longer be seen only as a remote, curiosity-powered generator of scientific discoveries that are then converted into new products and processes by applied research and development in the subsequent stages of technology transfer. This observation, however, only sets the stage for a more realistic account of the relationship between basic science and technological innovation.

Three questions of increasing importance arise about the dynamic form of the postwar paradigm. The least important is whether the neatly linear model gives too simple an account of the flows from science to technology. An irony of Bush's legacy is that this one-dimensional graphic image is one he himself almost certainly never entertained. An engineer with unparalleled experience in the applications of science, he was keenly aware of the complex and multiple pathways that lead from scientific discoveries to technological advances—and of the widely varied lags associated with these paths. The technological breakthroughs he helped foster during the war typically depended on knowledge from several, disparate branches of science. Nothing in Bush's report suggests that he endorsed the linear model as his own.[25]

The spokesmen of the scientific community who lent themselves to this oversimplification in the early postwar years may have felt that this was a small price to pay for being able to communicate these ideas to a policy community and broader public for whom science was always a remote and recondite world of affairs. This calculation may well have guided the draftsmen of the second annual report of the National Science Foundation as they stated the linear model in the simplistic language quoted earlier in this chapter. In any case, these spokesmen did their work well enough that the idea of an arrow running from basic to applied research and on to development and production or operations is still often thought to summarize the relationship of basic science to new technology. But it so evidently oversimplifies and distorts the underlying realities that it began to draw fire almost as soon as it was widely accepted.

Indeed, the linear model has been such an easy target that it has tended to draw fire away from two other, less simplistic misconceptions imbedded in the dynamic form of the postwar model. One of these was the assumption that most or all technological innovation is ultimately rooted in science. If Bush did not subscribe to a linear image of the relationship between science and technology, he *did* assert that scientific discoveries are the source of technological progress, however multiple and unevenly paced the pathways between the two may be. In his words,

> new products and new processes do not appear full-grown. They are founded on new principles and new conceptions, which in turn are painstakingly developed by research in the purest realms of science.[26]

Even if we allow for considerable time lags in the influence of "imbedded science" on technology, this view greatly overstates the role that science has played in technological change in any age. In every preceding century the idea that technology is science based would have been false. For most of human history, the practical arts have been perfected by " 'improvers' of technology," in Robert P. Multhauf's phrase, who knew no science and would not have been much helped by it if they had.[27] This situation changed only with the "second industrial revolution" at the end of the nineteenth century, as advances in physics led to electric power, advances in chemistry to the new chemical dyes, and advances in microbiology to dramatic improvements in public health. But a great deal of technological innovation, right down to the present day, has proceeded without the stimulus of advances in science. Chapter 2 reviews evidence that developments in military technology, an area in which America remained pre-eminent in the postwar decades, proceeded without much further input from basic science. And in recent decades, Japan has achieved its position in such markets as automobiles and consumer electronics less because of further applications of science than because of its thinking up better products and making good products better through small and rapid changes in the design and manufacturing process, which were guided by customer reaction and considerations of cost.[28]

But the deepest flaw in the dynamic form of the postwar paradigm is the premise that such flows as there may be between science and technology are uniformly one way, *from* scientific discovery *to* technological innovation; that is, that science is *exogenous* to technology, however multiple and indirect the connecting pathways may be. The annals of science suggest that this premise has always been false to the history of science and technology. There was indeed a notable *reverse* flow, from technology to science, from the time of Bacon to the second industrial revolution, with scientists modeling successful technology but doing little to improve it. Multhauf notes that the eighteenth-century physicists were "more often found endeavoring to explain the workings of some existing machine than suggesting improvements in it."[29] This other-way-round influence is called the oldest type of interaction of science and technology by Thomas S. Kuhn, who notes that Johannes Kepler helped invent the calculus of variations by studying the dimensions of wine casks without being able to tell their makers how to improve their already optimal design—and that Sadi Carnot took an important step toward thermodynamics by studying steam engines but found that engineering practice had anticipated the prescriptions from the theory he worked out.[30]

This situation was fundamentally altered from the time of the second industrial revolution, in two respects. One is that at least in selected areas, science was able to offer a good deal to technology, and this trend has accelerated in the twentieth century, with more and more technology that *is* science-based. But the other, complementary change, one that is much less widely recognized, is that developments in technology became a far more important source of the phenomena science undertook to explain. This was much more than a matter of instrumentation, which has loomed large in science at least since the time of Galileo. It was rather that many of the structures and processes that basic science explored were unveiled only by advances in technology; indeed, in some cases *existed only in* the technology. Hence, more and more *science* has become *technology* derived.

This development was illustrated by the research of Irving Langmuir on the surfaces of the devices being produced by GE and the other electronics firms of his day. It would not be right to say that the several-billion-year history of the universe had produced no analogs of the surfaces that so fascinated Langmuir. But neither humankind nor its scientific community had seen them until they were unveiled by the advancing technology of the electronics industry. By working out their physics, Langmuir earned a Nobel Prize in 1932 as he also cleared the way for significant further advances in the technology itself. In Leonard S. Reich's view, Langmuir felt that "understanding the principles of the physical world and making improvements to technology were part of the same venture" and that his "concern with applicability gave considerable direction to his research," influencing his choice of apparatus, analytical method, and conceptual outlook.[31] The developing technology of the electronics industry revealed the physical phenomena he probed, and his understanding of molecular interaction in crystals and surface films led to important advances in the technology.

A contemporary example of fundamental research that is technology based is provided by the work of the condensed-matter physicists who are seeking the fresh scientific knowledge that will allow semiconductors to be grown atomic layer by atomic layer. Although the knowledge laid down by the creators of solid-state physics between the wars was essential to understanding the transistor when it was discovered after World War II, what then transpired was more a triumph of technology than of science as the semiconductors moved through their successive generations, with astonishing reductions in scale and increases in speed. The miniaturization has now carried to the point where it may be possible to convey information by the location of individual electrons. But for this a fresh advance in fundamental knowledge will be needed—to see, for example, whether in circuits that consist of many quantum dots or wells an electron can behave simultaneously as a wave and particle, a finding that can be enormously important both for fundamental physics and for future technology.

The influence of technology on the course of basic science is clear in technological innovations in *processes* as well as products. This has characterized the role of medical practice in the advances of biological science. The evolving but incomplete technology of epidemic control in the nineteenth century influenced the use-inspired basic science of Pasteur. As Bruno Latour has shown, Pasteur lent a cutting edge to the broad public hygiene movement in France and Britain, whose calls to action had been frustratingly unconvincing before his discoveries on the sources of disease armed the movement with an adequate theory of the problem.[32] A further example described by Judith P. Swazey and Karen Reeds is the emergence of endocrinology from the work of clinical physicians concerned with the malfunction of particular glands.[33] In the latter part of the nineteenth century these physicians had observed a series of disorders such as diabetes, goiter, and cretinism that are now known to be glandular in origin. They connected their observation of these disorders with the anatomists' discovery of a series of ductless glands in the human body. Thomas Addison, the London physician who gave his name to Addison's disease, helped establish this link by recognizing that patients who had the symptoms of this disease also exhibited pathological changes in the adrenal glands. Another pioneer was the French physician Pierre Marie, who linked the appearance of the coarse and elongated features of acromegalic patients with pathological changes in their pituitary glands. In a similar way diabetes was linked to disorders of the pancreas—and myxedema and cretinism to disorders of the thyroid.

The research launched by these observations laid the foundations of the modern field of endocrinology, which has worked out the chemical regulation of physiological processes through the endocrine system. By the early twentieth century these studies had established that the ductless glands secreted directly into the blood-stream various hormones essential to the physiology of the body; the rival hypothesis that these organs detoxified the blood was decisively rejected. By the 1920s and 1930s this growing field had provided an understanding of the complex interactions of the several glands of the endocrine system; by the end of World War II, of the relationship between the endocrine and nervous systems. In recent decades attention has centered on the molecular processes by which cells and organs receive hormonal direction. Clinical observation of disturbances in the endocrine system and successful intervention in the process of disease have been as influential on research in the recent past as they were in the time of Addison and Marie. Pathologies have proved to be both a continuing source of insight into the system's normal functioning and a motive for extending basic knowledge.

Who Reaps the Technological Harvest from Science?

Experience also reveals as problematic the third element we have identified in Bush's conceptual system, the idea that a country can expect to capture the return in technology from its investment in basic science. A skeptic seated at Bush's elbow when he penned his claim that "a nation which depends upon others for its new basic scientific knowledge will be slow in its industrial progress and weak in its competitive position in world trade" might have pointed out that elsewhere *Science, the Endless Frontier* noted that the United States reached the front rank in industrial technology when it was still far behind Europe in basic science:

> In the nineteenth century, Yankee mechanical ingenuity, building largely upon the basic discoveries of European scientists, could greatly advance the technical arts.[34]

The question of who reaps the technological rewards from advances in basic science was scarcely asked in the postwar world, with the United States so in the ascendancy in both science and technology.

But the world could scarcely miss this lesson now that the Jap-

anese have all over again shown that the greatest strides in productive technology can be made by a country that is well behind
in basic science, with Japan cast in the technological role earlier
played by the United States and America cast in the scientific role
earlier played by the Europeans. Ralph E. Gomory and Roland
W. Schmitt have observed that the disparity between the Japanese
edge in technology and lag in science is more easily understood
when it is noted that the United States was the world leader in
technology by the 1920s, well before it became preeminent in
science.[35] Indeed, chapter 4 discusses the danger of the world's
*over*learning this lesson the second time around and the need to
prevent the concern for *relative* standing in productive technology
from eroding a collective commitment to renewing the world's
stock of scientific knowledge.

A Paradox in the History of Ideas

This canvass of the experience of science confronts us with a
notable puzzle of intellectual history. The annals of research so
often record scientific advances simultaneously driven by the quest
for understanding and considerations of use that one is increasingly led to ask how it came to be so widely believed that these
goals are inevitably in tension and that the categories of basic and
applied science are radically separate. Of course, a great deal of
research is dominated by only one or the other of these goals. Niels
Bohr, as he groped for a model of the structure of the atom early
in this century, brilliantly exemplified the scholar-scientist engaged
in a pure search for understanding. Equally, Thomas Edison, as
he drove his research team to complete the development of commercially marketable electric lighting, exemplified the applied investigator wholly uninterested in the deeper scientific implications
of his discoveries. Edison gave five years to creating his utility
empire, but no time at all to the basic physical phenomena underlying his emerging technology.[36] When others persuaded him that
his primitive grasp of these things would ultimately limit his engineering ventures, he, like the "half-educated electricians" who
bowed to the heirs of Maxwell in Britain, ran up the white flag of
surrender, confessed that he never had understood these matters,
and hired some technicians trained in Maxwellian field theory.[37]

Yet, however faithful these examples to the goals of a great many scientists, the annals of science are also rich with cases of research that is guided both by understanding and by use, confounding the view of basic and applied science as inherently separate realms.

We are therefore left with a paradox in the history of ideas. Although the vision of science and technology articulated after the war by Bush is by no means universally held, it has so permeated thinking about the scientific enterprise as to constitute a paradigm for understanding the relationship of science to technology in the late twentieth century. If this postwar paradigm is in tension with much of the experience of science, how can this view have become so prevalent? This puzzle is analyzed in chapter 2. As the paradox is resolved, we will gain a deeper understanding of the system of ideas that emerged from the war. The stage is also set for the effort made in later chapters to reshape the postwar paradigm and to explore what a new vision implies for policies on science and technology.

2 | THE RISE OF THE MODERN PARADIGM

IT WAS WIDELY ACCEPTED in the postwar years that basic science can serve as a pace-maker of technological progress only if it is insulated from thought of practical use. What a paradox in view of how often those who built modern science were directly influenced by applied goals as Louis Pasteur was influenced by practical goals throughout his fundamental work in microbiology. How is this paradox to be resolved? Why has this vision of science and its role in technological innovation, which is so evidently incomplete, prevailed?

Although the modern statement of this paradigm view was defined only after World War II, we must reach much farther into the past for the ideological and institutional insights that help to resolve this paradox. Indeed, the ideological sources of this view go back to the origins of the ideal of pure inquiry in the Greek world, although we owe to early modern Europe the corollary belief that such inquiry can improve mankind. The institutional influences on this vision lie in Europe and America in the nineteenth and twentieth centuries; the belief that basic and applied research are separate ventures was built into the institutional arrangements of science and technology in England and Germany in the nineteenth century and in America in the twentieth century. To these influences were added the political motives of the scientific community to accept a paradigm view that justified the government's continued support of basic science while restoring the autonomy of science that had existed prior to World War II. Many of the issues surrounding science and technology today can be clarified by tracing the rise of the paradigm view in the classical, European, and American experience and by examining the political context of science policy after World War II.

The Ideal of Pure Inquiry in Classical Times

We begin, as scientific philosophy did, with the Greeks. Although there is no Greek equivalent for the modern category of "science," it was clearly the Greeks who invented scientific inquiry. The earlier, technically advanced civilizations of the Egyptians, Assyrians, and Babylonians, and of the Indians and Chinese, failed to do so. The Greeks succeeded first of all by discovering "nature," although this has an odd ring to the modern ear. They were willing to regard the world as a natural system governed by general and discoverable natural causes; to leave the gods out, in Farrington's phrase.[1] To this they added the belief that these causes could be clarified by rational inquiry. The "Greek miracle," which began with the Milesian philosophers in the sixth and fifth centuries B.C., grew out of their beliefs in natural cause and the power of reasoned inquiry.

The Greeks could, in a sense, have invented these things *only* by severing the tie between philosophic inquiry and the practical arts. The technology of the earlier civilizations showed an awareness of some very general properties of natural things. Without a practical geometry the Egyptians could not have redrawn the land boundaries in the Nile Valley after the yearly flood. Without a practical astronomy the Babylonians could not have predicted eclipses of the sun and moon. But causes were left, in these civilizations, to the realm of the supernatural. A cuneiform text that survives from the Babylonians takes up the question of *why* the heavenly bodies should so precisely obey the empirical regularities their astronomers had worked out. The answer was that a group of gods had decreed that it be so—in one of the more famous committee decisions in the history of science. By freeing their natural philosophy from the practical arts and focusing it on the quest for general understanding, the Greeks transformed the practical geometry of the Egyptians into the works of Pythagoras and Euclid and arrived at explanations of the nature of matter (*physis*), such as the atomic theory of Democritus, that are hauntingly prescient of modern science, even if they penetrated these mysteries only to a limited degree.

This philosophic motive for severing inquiry from use was strongly reinforced in Greek civilization by the consignment of the practical arts to people of lesser station—and manual labor increasingly to slaves. As a result, as early as the Ionian philosophers practical utility was rejected as a legitimate end of natural philosophy, and this became a core belief in the Platonic and Aristotelian systems of thought. This denial took a double form in Plato. His ideal Republic radically separated those engaged in philosophic inquiry from those engaged in the manual arts, assigning a more exalted station to the former. And the ultimate reality sought by the philosopher lay in general forms or ideals rather than in the objects of the familiar world.[2]

Although Aristotle retreated from Plato's idealist philosophy and strongly encouraged empirical observation, his aim was still to identify the general in the particular, sifting his observation by the rational, deductive method employed. He continued to reject practical utility as the purpose of inquiry, sharing with his teacher the belief that philosophic investigation carries its own reward. Indeed, each made this a centerpiece of his psychology, arguing that philosophy is essential to happiness, since the reasoning faculty is the highest part of the soul.[3]

Aristotle's attitude toward the philosophic calling is revealed by his remarks in the *Metaphysics* that "as more arts were invented, and some were directed to the necessities of life, and others to recreation, the inventors of the latter were naturally always regarded as wiser than the inventors of the former, because their branches of knowledge did not aim at utility."[4] Geoffrey Ernest Richard Lloyd concludes, "Aristotle does not so much ignore the possibility of putting theoretical ideas to practical use, as positively glory in the ideal of the pursuit of knowledge for its own sake."[5] He quotes from the passage in the *Metaphysics* in which Aristotle

says that, since men "philosophized in order to escape from ignorance, evidently they were pursuing science in order to know,
and not for any utilitarian end."[6]

The bias against practical use in Plato and Aristotle continued
to dominate Greek thought throughout the Hellenistic and Greco-
Roman periods, despite the considerable Greek achievements in
engineering and the martial arts. Archimedes is a celebrated example. He was, by Plutarch's account, persuaded by King Hiero
to design various warmaking machines. But he thought so little of
this as a subject of philosophic discourse that he included not a
word about mechanics or engineering in the vast body of theoretical writings he left behind, to the evident satisfaction of the neo-
platonic Plutarch. In the words of Lloyd,

> the educational elite whom Plutarch typifies generally com-
> bined contempt for the life of the engineer with ignorance con-
> cerning his work. This attitude, which had the weighty support
> of Plato and Aristotle, is, without a doubt, the dominant one in
> writers of all periods in antiquity.[7]

A. C. Crombie judges that "it remained characteristic of Greek
scientific thought to be interested primarily in knowledge and understanding and only very secondarily in practical usefulness."[8] In
the twilight of the classical Greek world the gulf between inquiry
and use deepened still further as the Stoics, Epicureans, and Neo-
platonists came to think of philosophy less as a means of knowing
than as a means of achieving peace of mind in an increasingly
troubled world.

The major exceptions to all of this were the Hippocratic physicians. Their surviving writings, including detailed accounts of
clinical practice, show that from the time of Hippocrates in the
fifth century B.C. Greek medicine had inquired into a wide range
of human anatomy and physiology to develop the medical and
surgical means of dealing with wounds, fractures, and disease. The
Greek physicians added to the knowledge gained from surgical
practice what could be learned by dissecting animals and even, for
a brief period in Hellenistic times, by vivisecting condemned crim-
inals bound over to them by their Ptolemaic rulers.

The Hippocratics were in the mainstream of Greek philosophy in their attention to natural causes—again, having the courage to leave the gods out. But they were virtually alone in turning their inquiry and learning to the improvement of a practical art. By contrast, Aristotle, when he urged that animals be dissected as a means of extending biological knowledge, made clear that improving the practice of medicine was not at all what he had in mind.[9]

Greek science was the foundation upon which European science later built. By a remarkable channel virtually the whole of the natural philosophy of Plato and Aristotle, as well as of Archimedes and other writers from the Greek world of science, became available in the Latin West by the late Middle Ages. This did not result from the translation of these works into Latin in Roman times, since the Romans had only superficial interest in Greek science. Instead, the Greek scientific corpus was translated into Arabic after the Islamic conquest and spread to the great centers of Arab learning throughout the Mediterranean world. It was therefore found in Spain and Sicily when a resurgent Christian Europe reclaimed parts of Islam, although elements were recovered from Byzantium as well. It was then translated from Arabic in the twelfth and early thirteenth centuries by an extraordinary band of scholars who made the new learning available in the universal scholarly language of the Latin West. The arrival of these intellectual riches was a prime stimulus for the organization of the new universities in Oxford, Paris, Bologna, and Padua, which focused parts of their curricula on the new science.

The Ideal of the Control of Nature in Early Modern Science

The system of scientific thought that entered Europe in the thirteenth century came, in Crombie's words, "as a complete and for the most part coherent whole . . . a system of rational explanations in power and range quite beyond anything known earlier in the Latin West, and one the general principles of which in fact dominated European science until the seventeenth century."[10] It took a century for this system to be absorbed and a further century for the medieval scholastic science that was to be built on this new base to reach its full height. Without this absorption of classical science the scientific revolution of Galileo and Newton in the seventeenth century is scarcely imaginable.[11] Moreover, the natural philosophy of the Greek world, and the classical philosophy and literature that entered during the Renaissance, continued to be read by the classes in western Europe from which the natural philosophers of later centuries were largely drawn.

There is little doubt that it kept alive the view of the superiority of pure science that was so deeply rooted in the Greek world. And there is little doubt that the belief in knowledge for its own sake was echoed in the thinking of European scientists in every subsequent period. It was, for example, heard in our own day in the retrospective offered by C. P. Snow on the outlook of his fellow Cambridge scientists in his well-known analysis of the "two cultures" of the sciences and the arts:

> We prided ourselves that the science that we were doing could not, in any conceivable circumstances, have any practical use. The more firmly one could make the claim, the more superior one felt.[12]

But the belief in pure science in the Greek tradition was not the only strand of European thinking about science. There were distinctive elements of the European outlook, quite apart from the revolutionary changes in the substance of scientific theory, which ultimately overthrew the Aristotelian structure of scholastic science. As early as the thirteenth century a utilitarian case for science was made in terms that would have been quite alien from the classical world. Europe's natural philosophers were readier than the Greeks to see their science as a means of controlling, and not only of understanding, nature.

We have only partial clues as to why this should be true. Those performing the practical arts held a different station in late medieval and early modern Europe than they did in the Greek world. The medieval guilds lent these arts considerable prestige, and the Christian tradition gave manual labor a meaning quite different from the ancient world's. Roger Bacon and other clerics who brought Greek science to Christian Europe belonged to religious orders that affirmed the worth of manual service. And the Christian faith produced more complex ideas about the uses of science, such as the belief that scientific knowledge might allow mankind to regain the dominion lost in the Fall.

The coupling of knowledge and action characterized the Italian Renaissance, the cultural seed-bed of early modern science. Charles Coulston Gillispie asserts that

> the enterprises of a Brunelleschi, a Leonardo da Vinci, a Michelangelo, a Vasco da Gama, a Christopher Columbus [were] . . . animated by the same instinct that later formed a Galileo, namely, that knowledge finds its purpose in action and action its reason in knowledge.

For Gillispie these behavior patterns "were what made the culture of the Renaissance in Italy the matrix wherein ancient and scholastic learning and technique were converted into modern science and engineering"[13]—a transformation that might be thought to mock the term "Renaissance"—although Gillispie also credits the influence of Platonic idealism on Galileo's search for the general in the particular.

The greater interest of European scientists in the useful arts went hand-in-hand with their experimentalism. Indeed, the scientific breakthroughs of the seventeenth century were partly due to their readiness to apply to science techniques they borrowed from the arts and crafts. And they were equally ready to lend their talents to the improvement of technology—Tartaglia and Galileo to the improvement of military ordnance; Stevin to hydraulic engineering; Leibniz and Huygens to power machinery; Galileo, Torricelli, Descartes, Huygens, and Newton to the improvement of the telescope.[14]

The most influential spokesman for this utilitarian ideal was Francis Bacon, the English philosopher who is also remembered as a champion of inductive method. Bacon is the source of the aphorism that knowledge is power, by which he meant power over nature. In his view the purpose of science was mastery over nature, and as he prepared to displace the older, Aristotelian *Organum* with his *Novum Organum* he declared that knowledge and human power are synonymous, since the ignorance of the cause frustrates the effect, for nature is only subdued by submission.[15]

In view of the link between human knowledge and human power, Bacon believed that the utilitarian, material fruits of power would confirm the adequacy of the knowledge.

The modern distinction between science and technology was considerably blurred in the thought of Bacon and his contemporaries, who tended to merge the two. In Bacon's view, techniques *were* knowledge, rather than the fruits of knowledge. The outlook was one of an encyclopedic science that sought to establish where things *fit*, an exercise in systematics quite different from the more modern view of this relationship formed in the nineteenth century by the rise of positivism and the influence of German idealism and Kant's neo-Platonism. Indeed, the modern concept of "technology" was a neologism that came into common use only in the nineteenth century.

The influence of Bacon's utilitarian conception of science was greatly extended when it was written into the charter of the Royal Society, founded after his death. His vision of a research institute, Solomon's House, set out in his posthumously published *New Atlantis,* led to the society's formation. Its charter, granted in Restoration times, included good Baconian language charging the Fellows with "further promoting by the authority of experiments the sciences of natural things and of [the] useful arts . . . to . . . the advantage of the human race,"[16] and the society did undertake a variety of investigations into navigation, mining, and other practical technologies. Bacon's utilitarian views, as well as his inductive methods, greatly appealed to the eighteenth-century French encyclopedists and have remained an important part of Western thinking about science in all subsequent periods.

But the success of Bacon's vision is not without its ironies, since his view of the close link between science and practical technology outpaced reality by three hundred years. So general in our own day is the belief that science paves the way for technology that the actual history of this relationship may seem surprising. As noted in chapter 1, the link between scientific advance and technological innovation was in fact weak and problematic down to the time of the "second" industrial revolution in the late nineteenth century, when technical progress in chemical dyes and electric power and public health clearly depended on advances in chemistry, physics, and biology.[17]

Institutionalizing the Separation of Pure from Applied in Europe

The claims that practical use made on the decisions of science and the energies of scientists in the centuries from Bacon to Faraday were severely limited by the scant help the emerging sciences could offer to the flourishing European technologies of the time. In this long interval the increasing gap between understanding and use drained Bacon's utilitarian call to action of its immediacy and converted it into a faith that the endeavors of science would eventually improve the human condition. A natural philosopher who subscribed to this faith could work in his laboratory, heeding only the quest for understanding, while believing that his discoveries would somehow—at a later time and in other hands—benefit mankind.[18] Hence, the emergence of modern science before it could offer much help to the practical arts converted the Baconian tradition into an attenuated form in later centuries, the form incorporated in the canons of research articulated by Vannevar Bush in the mid-twentieth century. The strong Baconian doctrine set out in the *New Atlantis* and written into the charter of the Royal Society foresaw immediate benefits from the fusion of science and practical concerns, with the scientist directly linking the two. But the attenuated form of the Baconian tradition embraced by many of the natural philosophers of later centuries envisaged a deferred link between advances in science and improvements in human welfare. In this latter view, the two were separated in time and agency.

The separation in both terms was reinforced by the institutional arrangements of science and technology in this period, particularly by the difference in social background and economic circumstance of those who advanced science and the practical arts. For most of the eighteenth and nineteenth centuries the natural philosophers were, for compelling economic reasons, very different from those who were, in Multhauf's phrase, "'improvers' of technology."

Having little impact on technology, the science of the day brought little economic return. It could therefore be done only by people of means or patronage, including clergymen and certain other professionals.

By contrast, technology was in the hands of those who were engaged in practical work and sustained by its economic return. As in most technologically advancing societies, their small but important contributions went typically unsung, until a change in the patent laws permitted "inventors" to claim part of the return from technological change.[19] But the inventors of European society were far removed from the gentlemen scientists. Although their approaches to invention could be highly systematic, they had little theoretical grasp of science and needed little as the pace of their contributions accelerated into the industrial revolution.

All of this was evident in Britain in the eighteenth and nineteenth centuries. Many of the scientific advances of the period were the work of the wellborn or well-educated, some of whom remained in university settings, although scientists such as Joseph Priestly, Dalton, and Michael Faraday were of modest origin. Indeed, a tribute needs to be paid to the scientists of the Scottish Enlightenment, who helped make science respectable for the aristocratic classes in England. By contrast, the great technical advances of the industrial revolution were almost wholly the work of practical inventors and entrepreneurs, typically less educated and of lesser social station, who had and needed little theoretical understanding of science. In the words of Eric Ashby:

> The industrial revolution was accomplished by hard heads and clever fingers. Men like Bramah and Maudslay, Arkwright and Crompton, the Darbys of Coalbrookdale and Neilson of Glasgow, had no systematic education in science or technology. Britain's industrial strength lay in its amateurs and self-made men: the craftsman-inventor, the mill-owner, the iron-master. It was no accident that the Crystal Palace, that sparkling symbol of the supremacy of British technology, was designed by an amateur. In this rise of British industry the English universities played no part whatever, and the Scottish universities only a very small part; indeed formal education of any sort was a negligible factor in its success.[20]

With class an important element of the distinction, it was easy to think that scientific and industrial advances were in the hands of wholly different people of different backgrounds and training, who were attuned to different goals.[21]

Although this sense of root separation survived the nineteenth century, its institutional form was profoundly changed. As the century progressed, scientists were increasingly able to find support in the universities, and science became a more meritocratic calling. The creation of professional, economically viable research careers in the universities and research institutes powerfully stimulated the growth of science. It also institutionally reinforced the view that scientific inquiry should be pursued for its own sake; the nineteenth century fully reawakened this oldest Western outlook on the purpose of science. But the nineteenth century was also the time in which Bacon's proposed marriage of science and technology was at last consummated, and a number of leading scientists—with Kelvin a notable example—chose problems and pursued detailed lines of inquiry with an eye to practical technology as well as fundamental understanding.[22] The triumph of Maxwellian field theory over the more primitive ideas of the "half-educated electricians" was a watershed in the development of electric power in Britain and America.[23]

The technologist's role changed still more dramatically as science began to have a direct influence on technology. The spreading awareness that technical innovation would require the continuous application of scientific methods to industrial processes led to the creation of the technical schools, beginning with the *École Polytechnique* in France. By the end of the century it was clear that industry would employ large numbers of trained technologists or engineers, even if industry continued in many cases to prefer methods that were ruthlessly empirical.

It was the Germans who most fully institutionalized the new system. They did so first of all by making their universities an unparalleled setting for original scientific investigation, inspired by the great ideal of *Wissenschaft*. In the prior century the universities at Göttingen and Halle began to emphasize the creation of new knowledge through research. These stirrings were quickened when in the early nineteenth century new universities were founded in rapid order at Berlin (1810), Breslau (1811), Bonn (1818), and

Munich (1826) and the new emphasis spread to Leipzig and others of the established universities.[24] The result was an explosive release of scientific energy. Enrollments, staffs, and budgets soared. New disciplines and fields of specialization within existing disciplines were created. A strong link was forged between research and instruction as professors, who directed the work within their research institutes, also were responsible for teaching within these fields. New learning formats—specialist lectures, research seminars, laboratory experiences, monographic studies—were created to meet the needs of the scientific curriculum. There was a darker side to the German universities in this period—the rigidities of university structure that limited the development of important new fields, the professors' view of the research institutes as private baronies, the excessive career anxieties of staff who depended on the professors' autocratic authority, the single level of degree that provided more scientific knowledge than was needed by students headed for other careers but less than was needed by those aspiring to research careers. But the record as a whole was remarkable. It established the German-speaking universities of Germany, Austria, and Switzerland as the leading centers of scientific development.

The Germans used very different arrangements to support their rapid technological advances in the nineteenth century. They lodged applied science and development in the *Technische Hochschulen* and industry and imparted a new prestige to these technical schools and the careers to which they led. The Germans thereby institutionalized a strong sense of the separation of technology from the pure science lodged in the universities and research institutes. Students were streamed for these alternative career paths from an early age, and the technical schools and technological institutes produced the trained personnel needed by the parts of industry that were increasingly driven by science. To sustain the rapid technical advances in fields such as chemical dyes and pharmaceuticals, the Germans for the first time developed extensive programs of applied research. Hence, they institutionalized a sharp separation between pure and applied *research*—and not only between scientific research and practical invention—creating a relationship that was, as Multhauf notes, reminiscent "of the antagonistic philosophers and tradesmen of antiquity."[25]

These institutional arrangements, which appeared to cleanly

separate pure from applied science, concealed some complexities in the interplay of the research goals of understanding and use. Given science's increasing importance for technological change, there were inevitable examples of research undertaken in the universities with an eye to practical technology as well as fundamental understanding. The chemical industry recruited university-trained chemists to produce the new synthetic dyes, by batch-processing methods that replicated what they had known in their university settings.[26] There were also important examples of research outside the universities that merged these goals. In the later decades of the nineteenth century the research needs of industry led to the creation of the Kaiser Wilhelm Institutes, the forerunners of the Max Planck Institutes of today, as centers of use-inspired scientific research. Yet German institutions strongly encouraged the separation of pure from applied science, giving the Greek ideal of pure inquiry, reinforced by German idealist philosophy, a remarkable modern revival in the universities. Helmholtz articulated the spirit of this revival in his Academical Discourse, delivered at Heidelberg in 1862, in which he declared, in terms echoed by Vannevar Bush eight decades later, that "whoever, in the pursuit of science, seeks after immediate practical utility . . . will seek in vain."[27] Meanwhile applied science, enjoying a hitherto unparalleled prestige, helped drive industry's growth.

Institutionalizing the Separation of Pure from Applied in America

The spectacular achievements of the Germans both in pure and in applied science made their system extraordinarily influential. They were so excellent in each that their institutional arrangements were thought to be the natural order of things by an admiring world. Thousands of Americans flocked to the German universities in the late nineteenth and early twentieth centuries.[28] This tide, which was at flood in the 1880s, reflected the lack of an American system of advanced studies adequate to the needs of a rising industrial nation and was a standing challenge to create one. It is no surprise that the vision of scientific training and research brought back by these students played so large a role in the way this challenge was ultimately met, although the detailed path of institutional development in America would diverge from Germany's.

In view of the strongly utilitarian character of prior American science it is almost astonishing how successful the American emulators of the German experience were in providing pure science an institutional base within the new universities. From colonial times America's scientists had found it natural to mix the quest for understanding and use. Theirs was a technologically oriented society with a powerful belief in progress, and they were as likely to be drawn to their science by the belief it would help in practical ways as they were by the fascination of discovery. Before the Revolution, Benjamin Franklin wrote this fusion of goals into the charter of the American Philosophical Society. In 1819 Silliman dedicated the first *American Journal of Science* to the application of research "to the arts, and to every useful purpose."[29] By mid-century the scientific schools that flourished at Harvard, Yale, Princeton, and other colleges were amalgams of what today would be seen as pure science and engineering. In the decades after the Civil War the mission of the agricultural schools and experiment stations was an inherent meld of science and technology. In the latter part of the century the impressive scientific establishment within the federal government was imbued with a belief in the public benefits of science and prepared the ground for the vigorous conservation movement of the Progressive era.[30]

This same blend of goals is evident in the work of the handful of genuinely distinguished scientists produced by America in the eighteenth and nineteenth centuries. The immersion of Franklin in practical technology is too well known to require comment. The inventor of the Franklin stove was unlikely to put aside practical goals when he pursued his research in electricity, his most important scientific work. The legacy of this research included the supremely practical lightning rod, as well as a quantum leap in fundamental knowledge.

This blend of motives is equally clear in the work that led Joseph Henry to discover, independently of Faraday, the principle of electromagnetic induction, converting magnetism to electricity and preparing the way for the generation of electric power. In his earlier explorations of steam Henry was aware of the spreading use of steam power in industry, and talk of movable "steam wagons" was already in the air. When he turned to electromagnetism he well understood the practical benefits that could flow from significant discovery. He described his first electromagnetic machine, capable of reciprocating motion seventy-five times a minute, as "a philosophical toy" but noted that a modified version "may hereafter be applied to some useful purpose." This device in fact anticipated the direct-current electric motor.[31] His decision not to press ahead was based on his calculation of the relative efficiency of generating steam power from coal and electric power from the zinc required for batteries. Henry eschewed commercial invention and made no effort to become one of America's celebrated nineteenth-century inventors. Yet he found it entirely natural to allow calculations of use to influence the choices by which his science unfolded.

The case of Josiah Willard Gibbs, the founder of statistical mechanics and formulator of the Phase Rule of chemical thermodynamics, is more elusive. The intensity with which this most modest of scientists pursued the implications of the second law of thermodynamics, and the mathematical elegance of his methods, might suggest a pure quest for understanding, and so might the fact that he was the third person to earn Yale's new Ph.D. degree. But Gibbs chose his problem in an age of steam and electricity, and he understood the enthusiasm of his Yale masters for appointing him to a chair in physics, the science that seemed capable of providing a more general understanding of these intensely practical phenomena. His treatment of equilibrium in heterogeneous substances was so elegant that its practical worth was for some years lost on industry, despite repeated plugs by his English admirer James Clerk Maxwell, so few were his countrymen who were capable of reading his work. A number of subsequent investigators independently discovered parts of what Gibbs had found—only to realize how much more general and powerful a formulation he had given. In the end, his work was seen in both Europe and America as having the practical value that the ethos of his time always led him to expect it would.

The utilitarian ethos of American science was profoundly changed by the rise of the research universities. Their advent is a remarkable chapter in American education. The need to create institutions of advanced learning became increasingly clear as thousands of students who had earned college degrees went on to study at the German universities in the late nineteenth century. The efforts to fill this gap in American higher education were generously supported by America's economic expansion, particularly by the private individuals who acquired great wealth in the decades after the Civil War. And the way was led by a remarkable group of educational entrepreneurs—James Rowland Angell, Charles W. Eliot, Daniel Coit Gilman, G. Stanley Hall, William Rainey Harper, David Starr Jordan, and others—many of whom gained a vision of what might be done from their studies in the German universities.

The wave of change that swept over American higher education created new universities—Cornell in 1865, Johns Hopkins in 1876, Clark in 1887, Stanford in 1891, and Chicago in 1892. It created graduate schools and Ph.D. programs that transformed private colleges such as Yale, Harvard, and Princeton and state colleges such as Michigan, Wisconsin, and California at Berkeley into genuine universities. It spread the elective system from Eliot's Harvard to other institutions, opening up the curriculum to the new fields of study.

These changes gave strong institutional support to the vision of the universities as centers of original research and teaching in pure science. This point is strikingly illustrated by the contrast between the freedom to pursue research enjoyed by a university-based scientist in the late nineteenth or early twentieth century and the situation of Joseph Henry, whose grinding responsibilities to teach in the Albany Academy delayed the report of his discovery of electromagnetic induction until Faraday was already in print. Gillispie has remarked that the first trait of a profession is the custody and development of a body of knowledge and that the second is the provision of economically viable careers.[32] George H. Daniels has shown that even before the Civil War American scientists had gained control of their increasingly esoteric fields and shut out the amateurs.[33] But it was the American research universities, like their German inspirations before them, that converted original scientific research into an economically viable professional career.

Indeed, in some respects their plan improved on the German original. One was a clearer, institutionalized way of entering the newly defined careers of university research. In Germany, well into the nineteenth century, the capacity to do distinguished research was still regarded as a charismatic trait possessed by certain gifted individuals, rather than a trained capacity, and it was in the interest of the holders of research chairs, who *were* the university in a corporate sense, to sustain this view. As a result, the way individuals entered research careers was never clearly spelled out, to the grief of several generations of *Privatdozenten* who waited in the shadows for their own extraordinary gifts to be recognized. The American universities took a more straightforward approach to all this. Rather than clinging to a single level of degree, they enrolled as graduate students those who had finished their baccalaureate studies and now wanted to be trained for research careers. The most proficient of the Ph.D. recipients were then appointed to junior faculty positions in the research universities—where, if they realized their potential, they rose to full professorial rank.[34]

The American plan also improved on the German original by vesting authority over a field in an egalitarian department of peers and not in a single professor, holding *the* chair. The German system fell victim to the very research energy it released, since the rapid specialization and changing content of an exploding body of knowledge tended to overwhelm the capacities of a single professor, however distinguished. These changes were far more easily accommodated by the American department of peers, chaired by a colleague whose role was seen as administrative.

Understandably, these institutional arrangements for basic research tended also to develop a view of pure science quite different from the American outlook earlier in the nineteenth century. The promoters of these arrangements could hardly import from Germany the idea of a career in scientific teaching and original research without also importing the ideal of knowledge for its own sake and the distinction between pure and applied science. Some degree of acceptance of these things was needed to establish basic science in the universities.

As pure science was being provided with an institutional home in the universities, the sense of separation of pure from applied was heightened by the institutionalization of applied science in industry. In America too, industry mounted substantial applied *research* that went beyond at-the-bench improvements in technology as the content and methods of science were increasingly brought to bear on industrial processes. In the larger and more complex industrial firms untrained empirical technologists were supplemented by applied scientists and engineers, many trained in the new colleges of engineering and the polytechnical institutes. The way was led by the chemical industry. The electrical industry clung to empirical methods for several decades but fell in line with the rise of twentieth-century electronics. As other industries followed suit, applied industrial research grew in range and volume, making American science and technology, in Edwin Layton's phrase, "mirror-image" twins:

> The technological community, which in 1800 had been a craft affair little changed since the Middle Ages, was reconstructed as a mirror-image twin of the scientific community. . . . In place of oral traditions passed from master to apprentice, the new technologist substituted a college education, a professional organization, and a technical literature patterned on that of science. . . . As a result, by the end of the 19th century technological problems could be treated as scientific ones; traditional methods and cut-and-try empiricism could be supplemented by powerful tools borrowed from science.[35]

These mirror-image twins, so alike yet different, institutionalized the sense of distance between basic and applied that was welcomed by the pure-science fields within the universities.

The role claimed here for institutional factors in reinforcing the separation of pure from applied in Germany and America extends an observation by Thomas S. Kuhn, after Multhauf. Multhauf saw the separation as rooted in a radical difference of temperament or outlook between scientists and improvers of technology, concluding from his wonderfully learned comparison over two millennia that they are simply "different species, interdependent and even occasionally transmutable, but persistently distinct, like land- and water-dwelling creatures."[36] Kuhn, in an essay of equal historical sweep, draws a similar contrast but sees this difference as characterizing whole cultures or societies, "for almost no historical society has managed successfully to nurture both [science and technology] at the same time"—the Germans being a revealing exception:

Greece, when it came to value its science, viewed technology as a finished heritage from its ancient gods; Rome, on the other hand, famous for its technology, produced no notable science. . . . Britain, though it produced a significant series of isolated innovators, was generally backward in at least the abstract and developed sciences during the century which embraces the Industrial Revolution, while technologically second-rate France was the world's preeminent scientific power. With the possible exceptions (it is too early to be sure) of the United States and the Soviet Union since about 1930, Germany during the century before World War II is the only nation that has managed simultaneously to support first-rate traditions in *both* science and technology. Institutional separation—the universities for *Wissenschaft* and the Technische Hochschulen for industry and the crafts—is a likely cause of that unique success.[37]

The German exception leads Kuhn to offer an institutional rather than a societal explanation of the separation. In his view, the Germans recognized that science and technology advance by inherently different processes and created distinct institutional settings in which each could thrive. It is the more remarkable that the Germans institutionalized a sharp separation of basic science from applied science and technology in the century that finally harnessed technological to scientific progress. They did so by lodging pure science in the universities and research institutes and technology in the technical high schools and industry, drawing the distinction between pure and applied science exogenously to the universities.

Institutional factors, no less in America than Germany, helped create the perception that basic and applied science are separate ventures, pursued by distinct sets of people with distinct goals. But the institutional development of pure and applied science in America differed in notable respects from the prior German model. The American innovators were unwilling to dedicate their universities only to pure science, even if they had been able to do so. Supported by the new captains of industry and working within a pragmatic society with a Baconian tradition of science, the organizers of the research universities made a place for applied fields that were of interest to vari‾ ..; of their constituencies. One of these was agricultural science, strongly supported by the Morrill and Hatch Acts, which helped create a remarkable system of land-grant colleges, agricultural experiment stations, county agents, and extension services.[38] Another was biomedicine, as the new, scientifically aspiring medical schools were organized. Yet another were the engineering fields that were heirs to the scientific schools of the mid-nineteenth century, a development strongly reinforced by such free-standing engineering schools as the United States Military Academy, the Massachusetts Institute of Technology, Lehigh, and Rensselaer Polytechnic Institute. Although the engineering schools and departments seemed to reinstitutionalize in the American setting the empirical separation of applied from pure physical science, there was no drive to expel engineering from the emerging universities. The distinction between basic and applied science was simply drawn *within* the universities.

But a subtler reading can be offered of this aspect of the American experience. The applied fields, while they seemed to repeat the separation of basic from applied science, have in fact provided an institutional home for research that is driven by the goals of understanding *and* use. Similarly, institutions outside the universities, with Bell Labs the prototype, provided a home for research melding these goals.

Reinforcing the Separation:
The Aftermath of World War II

It is difficult to exaggerate how profoundly the relationship between science and government was transformed by World War II. The federal government had supported scientific activites from the beginning of the republic, and by the second half of the nineteenth century a substantial part of the science being done in the United States was in the hands of such federal establishments as the Smithsonian Institution and U.S. Geological Survey, together with the agricultural experiment stations established with federal support. But the German model of advanced scientific studies spread to America via the nascent research universities, and these institutions laid the groundwork for their preeminent scientific role in the twentieth century mainly with funds from private donors, philanthrophic foundations, state legislatures, and fee-paying students. Indeed, by the interwar years the academic scientists held a deeply rooted hostility toward the idea of federal support, out of concern over the control of the content of research such support might bring, the fear that science might lose, in a word, its "autonomy."

All of this was overthrown by the war. Its scientific effort was in the hands of enlightened scientists, foremost of whom was Vannevar Bush, who recruited a battalion of gifted colleagues for the research tasks of the war, with the backing of the strongest president of the twentieth century. Bush's Office of Scientific Research and Development became, as A. Hunter Dupree has said, "the nearest thing to a true central science organization in all of American history."[39] An unparalleled flow of resources funded projects in basic science, including the basic nuclear research leading to the weapon that altered the final stage of the conflict. As the war drew to a close, the scientific and policy communities were agreed that the federal investment in science should continue, and at Franklin D. Roosevelt's request, a request at least partially inspired by Bush himself, Bush prepared a plan for sustaining this investment in peacetime.

The personal prelude to the Bush report was Bush's own role in bringing America's scientific strength to bear in the war, by a route thoroughly in keeping with Franklin Roosevelt's fondness for initiatives outside of the regular government structure. As the war clouds gathered over Europe, Bush, then president of the Carnegie Institution of Washington and formerly vice president of the Massachusetts Institute of Technology, and four other senior figures of the scientific community[40] held a series of discussions of what lay ahead. They shared Roosevelt's awareness of America's stake in the impending war. The international character of science made them acutely aware of the tragedy that had befallen many of Europe's most gifted scientists who had been driven from their posts by the totalitarian regimes.[41] And Bush and his colleagues understood better than Roosevelt that the coming war would be partly a scientific and technological conflict, in which the United States faced an adversary with formidable credentials in both. They felt that no time should be lost in tapping America's scientific potential.

Alone of the five, Bush was Washington based and, by chairing the National Advisory Committee for Aeronautics, had experience in linking science and engineering to defense needs. He also had the priceless asset of access to the president through his unlikely friendship with Roosevelt's legendary assistant, Harry Hopkins. In June of 1940 Bush laid before Roosevelt a one-page proposal asking the president to create a National Defense Research Committee that could begin to enlist the country's scientific resources for the coming conflict. Roosevelt issued the necessary executive order. Bush was named chairman and assured of access to the president.[42]

A year later NDRC became part of the Office of Scientific Research and Development in the Executive Office of the President, with responsibility for medical research as well. OSRD to a remarkable degree succeeded in bringing the nation's strength in science and industrial engineering to bear in the war. Bush and his associates had the confidence of the most influential parts of the scientific community and recruited many of the country's leading scientists and engineers for the war effort. OSRD contracted for this work, rather than operating research laboratories of its own, and pioneered the idea of contracting directly with the universities and industry, rather than with individual scientists. It also pioneered the idea of compensating for the full cost of the work, including indirect costs, establishing the principle of "no loss, no gain" for nonprofit institutions. Beyond this, it improved the incentives for industry by allowing those performing the research to retain any patent rights. The budgets that supported the rapid buildup of the nation's research effort, although a tiny part of the full cost of the war, were huge by the standard of prior federal outlays for science. Long before the war's end, thought was given to how the momentum of the wartime effort could be continued in peacetime.[43]

So impressive was the wartime role of science, capped by the making of the atomic bomb, that it is easy to miss the controversies surrounding the Office of Scientific Research and Development.[44] The Bush leadership was drawn from the scientific elite of the universities and industry, a fact typified by the identities of those who originally proposed the formation of the National Defense Research Committee. As they recruited the leading sectors of American science and technology for the war effort, they inevitably

inspired criticism from those who saw their generally elitist venture
for what it was.

This reaction was set against a backdrop of liberal and populist
grievance from the years of the Great Depression.[45] There was an
undercurrent of feeling in the 1930s that science somehow shared
with business the responsibility for America's economic collapse.
The belief that industry was controlling markets by patents on the
fruits of its research laboratories was fed by charges early in the
war that prewar patent pools linking American and German firms
contributed to U.S. technological unpreparedness for the war.
Roosevelt's own willingness to substitute "Dr. Win-the-War" for
"Dr. New Deal" was symbolized by Hopkins's continuing assistance to Bush's efforts. But the Bureau of the Budget was critical
of Bush's practice of compensating the universities and industry
for the indirect as well as direct costs of research. Complaints were
heard from inventors whose ideas for improving America's weapons were shunted aside.[46] And there was unease about concen-
trating OSRD's contracts in the country's leading universities and
in firms that might use the resulting patent rights to control their
markets.

Much of this criticism found its champion in Harley M. Kilgore,
a populist U.S. senator from West Virginia. In 1942 Kilgore introduced the first of a series of wartime proposals to remedy these
reported difficulties, and as the war progressed he increasingly
gave attention to the postwar role of the federal government in
science and technology. He was the first to call, in 1944, for the
creation of a National Science Foundation to promote basic and
applied research and scientific training after the war.

Bush correctly saw that the Kilgore proposals, whatever their
merits, would deliver the government's future role in science into
the hands of an agency in which the scientific community did not
have the leading voice, an agency that would encroach on the
autonomy of science that Bush had striven to protect even under
the severe constraints of the war. He therefore welcomed the idea,
first suggested by Oscar S. Cox, a lawyer in the Roosevelt administration, that the president should ask him to develop an alternative proposal for the government's peacetime role in science.
Roosevelt made his request on November 17, 1944. Bush's re-
sponding report, *Science, the Endless Frontier*, was sent to Presi-

dent Harry S. Truman in early July of 1945, as America's entry into the atomic age was about to dramatize the ascendant importance of science in the war.

In the broadest terms, the task Bush and his advisers set for themselves was to find a way to continue federal support of basic science while drastically curtailing the government's control of the performance of research. The war had induced many of the country's leading scientists to set aside their historic suspicion of government support and to lend their efforts to the scientific goals identified by OSRD. This was symbolized by the willingness of James B. Conant and Karl T. Compton, the presidents of Harvard and MIT, to serve as Bush's deputies during the war, although each of them would later decline to be the first director of the National Science Foundation. Countless scientists in university or industry laboratories let OSRD set their priorities and coordinate their efforts while the war was on, although Bush skillfully sought to give full scope to their creative imagination as they went to work on the country's military needs. But the war was over. The scientific community knew that a great opportunity would be lost if government funding was not continued. But it wanted the funding without a continuation of anything like the same degree of governmental control. It wanted, in other words, to restore the autonomy of science.

The prime means by which Bush and his advisers sought to ensure this autonomy was organizational. From one of the background panels formed to advise him on the answers to the four questions Roosevelt asked,[47] Bush absorbed the idea of creating a National Research Foundation with responsibilities in basic science as broad as OSRD's during the war. The Foundation was to be governed by a board drawn from the scientific community. The director would be chosen by the board rather than appointed by the president and confirmed by the Senate. The background panel, chaired by Isaiah Bowman, president of Johns Hopkins, went so far as to recommend that the Foundation be insulated from the federal budget process by being provisioned with an expendable endowment that would need to be renewed only at widely spaced intervals.[48] Bush pulled back from this extreme in his own report and restored the Foundation to the federal budgetary process. But there is no doubt that his proposed National Research Foundation, if it had been created, would have protected the autonomy of science by delivering the government's funding of basic science into the hands of an agency in which the scientific community had the leading voice.

Yet Bush did not entrust his strategic ends to organizational means alone. Much of the later significance of his report lay in the fact that he reinforced his case for government support of an autonomous scientific community by setting out a distinctive vision of the nature of basic science and its relationship to technological innovation, a paradigm view that became increasingly important as the plan for a National Research Foundation foundered. As already noted, Bush's first canon about basic research—that it is performed without thought of practical ends—was admirably designed to persuade the country and the policy community that attempts to constrain the free creativity of the basic scientist would be inherently self-defeating. His second canon—that basic research is the pacemaker of technological improvement—was designed to persuade the policy community that the investment in basic science would yield technology to meet a broad spectrum of the country's needs.

The Response to Bush's Plan

The reception of *Science, the Endless Frontier* was filled with irony, since Bush's organizational plan was defeated while his ideology triumphed. His plan for a largely self-governing National Research Foundation as broad in scope as his wartime Office of Scientific Research and Development was shattered by the policy process during the five-year interval between publication of his report in 1945 and the creation of the National Science Foundation in 1950. Although Bush's plan was immediately put in bill form by Senator Warren G. Magnuson of Washington state, resistance to its provisions by the president and Congress delayed its enactment for half a decade. The most important grounds of resistance were exactly the provisions by which Bush sought to insulate the new agency from political control. President Truman and his advisers never wavered from the belief that the agency must be administered by a director chosen by the president and confirmed by the Senate. Truman vetoed a bill that lacked this provision in 1947 when the Republicans controlled both houses of Congress, but Congress could not muster the votes to override the veto. The bill ultimately signed into law in 1950, after the Democrats regained control of Congress, required the director to be appointed by the president and confirmed by the Senate, although the National Science Foundation's first director, Alan T. Waterman, deferred on policy to the National Science Board, a part-time body generally conforming to Bush's plan.[49]

Bush's proposal also faced opposition in Congress from those who correctly saw the provisions for the agency's governing board as an attempt to give the country's leading scientists a dominant voice and wished to diversify the government's investment in science beyond the elite universities and firms. Many of Bush's critics in Congress, a supremely geographic institution attuned to constituency interests, equated a more egalitarian approach with a geographic formula for the distribution of funds.

There was also resistance to Bush's proposal on two other points. One concerned the rights to patents for developments flowing from government-sponsored research, a sensitive matter in the years before and during World War II. But this issue lost much of its charge when it became clear that the grants made by the National Science Foundation, as it was eventually named, would be largely for research in the universities rather than industry. A second point had to do with the social sciences. Bush and his allies were conservative on this issue and reluctant to embroil science in the value controversies that seemed endemic to the social studies. The bills inspired by Bush's report grouped social science with "other sciences," whereas those in Congress who wanted a more progressive role for science pressed for explicit mention of the social sciences. Bush's view prevailed in 1950, and only in 1968 was the law amended to specifically authorize the Foundation to support social science research.

During the five-year delay between the publication of *Science, the Endless Frontier* and the creation of the National Science Foundation the policy process fragmented the portfolio that Bush had sought to deliver into the hands of a National Research Foundation. The first part to go was nuclear research, which was vested by the Atomic Energy Act of 1946 in a five-person Atomic Energy Commission. Although a more general scientific agency was likely to have devolved direct responsibility for nuclear energy, as OSRD devolved responsibility for the work of its Uranium Committee to the Manhattan Project in 1943, the agency Bush envisaged would have had greater involvement in this aspect of the government's support of science. The law creating the National Science Foundation turned the tables, however, and required Atomic Energy Commission approval of NSF grants for research in nuclear science.[50]

The military services filled a second part of the vacuum left by OSRD with programs of their own for supporting basic research, a transfer that was cemented by the creation of the Department of Defense in 1949. Knowing that OSRD would soon be gone, the secretaries of War and Navy had earlier created a Joint Research and Development Board chaired, ironically, by Bush himself. The Office of Naval Research, created in 1945, launched a vigorous program of support for basic science, becoming in some respects a "National-Science-Foundation-in-waiting." Its National Research Advisory Committee was an important link with the university community and developed a successful early model of the peer-review system for selecting projects deserving support. Alan T. Waterman, ONR's deputy director and chief scientist, later became the founding director of the National Science Foundation.

Perhaps most tellingly of all, the proposed domain of a National Research Foundation was further diminished by the provisions made for biomedical research in the absence of a successor to OSRD. With Congress and the president deadlocked on the new agency, OSRD's Medical Committee arranged in 1947 to transfer its research contracts to the U.S. Public Health Service. The National Institute of Health, which had been largely an in-house laboratory for biomedical research since the 1930s, was renamed the National Institutes of Health and given a flourishing program of extramural grants, on which Congress showered appropriations on an ever-increasing scale.[51]

We may wonder how effective a general basic science agency governed by a board drawn from the scientific community could ever have been in the decades after the war. The scope of the proposed agency would have involved it in the affairs of a wide range of government departments. To cope with the pressures of a revived budgetary process, it would have needed powerful new

allies in Congress. It would also have needed the strong presidential backing that Bush found essential even in the war, when he
had functioned as the president's science adviser. White House
support was, for example, as essential when the scientific work on
the atomic bomb was transferred to OSRD early in the war as it
was when the development of the bomb was spun off to the U.S.
Army's Manhattan Project in 1943—with Bush continuing to
chair a "top policy committee" that reported directly to the president. With Roosevelt and the invaluable Hopkins gone from the
White House, Bush's ties with Truman and his aides were far more
distant. His relations with John R. Steelman, a presidential assistant with increasing responsibility for science policy, were particularly edged by tension, and Steelman won the president's blessing
for the idea of producing, within two years of Bush's report, his
own "Steelman Report," *Science and Public Policy*, which sought
in five volumes to supersede *Science, the Endless Frontier* in key
respects.[52] It is hardly believable that without strong backing from
the president and the centers of power in Congress, the authority
of an autonomous Foundation, dominated by the nation's scientific establishment, could have been nearly as broad as OSRD's
during the war.

In any case, the organizational plan set out by *Science, the Endless Frontier* was never put to the test, and the irony of its reception
is deepened by the fact that the defeat of the plan made it *more*
likely that its paradigm view of science and technology would
triumph. Indeed, if Bush's organizational plan had succeeded, it is
unlikely that his report would have left as deep an ideological
imprint as it did. Its defeat allowed the paradigm view to be
warmly endorsed by the R&D-intensive mission agencies as well
as by the National Science Foundation when it was eventually
created. Once the responsibilities for the support of basic science
were fragmented and its own scientific turf secured, the Defense
Department could endorse Bush's outlook as a way of cementing
its relations with the scientific community. An enterprising re-
porter for *Fortune* magazine went to a meeting of the American
Physical Society in spring 1948 and found that nearly 80 percent
of the papers were supported by the Office of Naval Research.[53]
Harvey Brooks and others have reminded us how far the onset of
the cold war restored the status quo ante of World War II for parts

of the scientific community. Indeed, that a reduced form of Bush's proposal did not wither and die during this five-year delay is due partly to how well the cold war sustained the belief that a strong scientific base was a key to military security.[54] The outbreak of the Korean War and the very real possibility of the cold war's turning hot was the urgent backdrop of the actual organization of the National Science Foundation at the beginning of the 1950s.[55]

The National Science Foundation was free to wholeheartedly endorse Bush's vision when it was created at the end of this five-year pause. With the narrower mission of supporting basic research in the universities, the Foundation could be expected to find the idea of pure research as the ultimate font of technological progress congenial, and the description of the "technological sequence" in the second annual report of the National Science Board, quoted in chapter 1, is as clear a statement of the linear model of technology transfer as can be found in print. Waterman, knowing how protective of turf the departments and agencies were, steadily resisted the Budget Bureau's urging that NSF take responsibility for thinking about science in the government as a whole.

The postwar event that made clear how deeply Bush's paradigm view had soaked into the consciousness of the policy community was the response to the launching of *Sputnik* in 1957. The United States might have seen *Sputnik* as a challenge to a particular sector of American technology and renewed its effort to match or surpass the Soviet achievement in this sector by building bigger booster rockets and perfecting other parts of its space technology. It did do this and ultimately put a man on the moon. But the policy community also responded to *Sputnik* as a general Soviet challenge to American *science*, since they were persuaded that breakthroughs in technology must be based on prior breakthroughs in science. President Dwight D. Eisenhower himself played an important role in this response. He installed James R. Killian of MIT as the first science adviser to the president since the days of Bush, with instructions to help plan the scientific response to *Sputnik*. He also created the President's Science Advisory Committee (PSAC), with Killian as chairman. But Congress was equally determined to meet this challenge to American science. As a result, the years after this unwelcome Soviet surprise were ones of soaring budgets for virtually every branch of American science—so that the innovations coming out of the other end of the pipeline leading from basic scientific advances to new technology would be *our* surprises and not theirs. From the launching of *Sputnik* to the landing of an American on the moon, federal support for *basic science* increased in real terms almost by a factor of five.

The mood generated by *Sputnik* and the early cold war challenge did not survive the 1960s. Vietnam, the turbulence on the campuses and in the cities, the end of the long years of economic expansion without inflation, the concern with harmful effects of technology—these and other factors sharply reduced the willingness of the policy community to go on funding the expansion of basic science that was so essential an element of the postwar compact between science and government.[56] Yet it was remarkable how effectively Bush's paradigm continued to supply the concepts for the debate between those who sought to change and those who defended the postwar bargain.

The pervasiveness of these concepts is shown by the exchange following the Defense Department's shot across the bow of the scientific community when it mounted a retrospective survey of the contribution of research to a series of current weapons systems. The results of this "Project Hindsight" undercut any simple belief that scientific discoveries were the immediate source of continuing improvements in military technology. Of the several hundred critical "events" in the development of twenty weapons systems, fewer than 1 in 10 could be traced to research of any kind and fewer than 1 in 100 to basic research untargeted on defense needs. Most improvements in this weaponry were found to be modifications of existing technology or the result of development activities inspired not by research but by an awareness of the technical limitations of existing systems.[57] In view of the role of science-inspired weapons in bringing World War II to an end and creating the postwar strategic balance, no one could seriously have thought that only a tenth or hundredth part of the country's military strength rested on prior basic research, although this study was a reminder of how many of the advances in technologies where the U.S. remained ascendant in a high-tech age were extensions of existing technology rather than the result of scientific discoveries, much as Multhauf would have led us to expect.

The scientific community reacted with understandable dismay, since Project Hindsight so directly challenged the value of the $400 million the Defense Department and other mission agencies of the federal government were annually investing in basic research. The National Science Foundation commissioned a rebuttal study, TRACES, that charted the antecedents of five technological innovations—videotape recorders, oral contraceptives, electron microscopes, magnetic ferrites, and matrix insulation. Since these were known to have sources in basic science, the rebuttal merely demonstrated what was already clear— that technological innovation *could* be science-based—rather than more generally assessing the dependence of new technology on advances in basic science.[58]

It is possible to criticize both the Hindsight and TRACES studies as searching for the sources of innovation in discrete "events" in science. Yet what is most notable about this exchange for our conceptual account is how deeply the outlook of each side was shaped by the postwar paradigm. In particular, both the Defense Department and the Science Foundation continued to think in terms of the linear model, with analysts from the Defense Department claiming that all that mattered for these defense technologies was the limited segment of this sequence from development to production and operations, while NSF's analysts argued the primacy of pure research in the rise of the technologies explored by the TRACES study. Neither side gave any real attention to the possibility that technological needs might inspire research of fundamental scientific importance. This is the more striking in the case of those who touted the TRACES findings, since there is clear evidence that some of the basic scientific research examined by this study was directly inspired by the technological uses to which it ultimately led.

Therefore, factors both ancient and modern help to resolve the paradox of why a paradigm view of science and technology so far removed from their true relationship came to prevail. The ideal of pure inquiry that underlay the first of Bush's canons is as old as classical antiquity; in the Greek world, only the Hippocratics stood out against the separation of philosophic inquiry from the applied arts. The very different attitude toward practical ends by Bacon's time linked understanding and use in a common framework of encyclopedic science. But as a deeper science, distinct from technology, emerged in later centuries, the natural philosophers came to believe that their discoveries would lead to the betterment of mankind but that this would be the work of others at a later time—a view that foreshadowed the second of Bush's canons. And by the time, three centuries after Francis Bacon, when this deeper science was demonstrably the source of new technology, the interactive ties of science and technology through use-inspired basic research were shrouded by a further historical accident in the mid-twentieth century—the effort of the scientific community to preserve the autonomy of publicly funded science by declaring that efforts to constrain the creativity of basic research by considerations of use were inherently self-defeating. With the triumph of the outlook of *Science, the Endless Frontier* the paradox was complete. But the tension between this paradigm and the actual experience of science remained, and the challenges to Bush's canons became more insistent as the country's needs shifted from the military to the economic sphere. The next chapter develops a view of this relationship more in keeping with the realities of basic science and technological innovation.

3 | TRANSFORMING THE PARADIGM

Half a century has passed since Vannevar Bush articulated the paradigm view of basic science and its role in technological innovation that was absorbed into the thinking of the scientific and policy communities after World War II. This framework of understanding, partly inspired by the ideal of pure inquiry in Western scientific philosophy and reinforced by the institutional separation of pure from applied science and by the postwar interests of the scientific community, has influenced science and technology policy over much of the succeeding period.

Yet this framework has come under heavy pressure as the policies to which it led seem less adequate for the needs of a different era. Indeed, these doubts have appeared in each of the major industrial countries. It is no longer believed that a heavy investment in pure, curiosity-driven basic science will by itself guarantee the technology required to compete in the world economy and meet a full spectrum of other societal needs. Britain, for example, issued a May 1993 White Paper on science and technology policy which flatly stated, "The Government does not believe that it is good enough simply to trust to the automatic emergence of applicable results [from basic research] which industry then uses."[1] In each of the industrial countries interest in harnessing science for the technological race is increasing, and this interest helps to create a climate that is receptive to a fundamental critique of the postwar framework for thinking about science and technology.

Early Dissents

Once the prevailing paradigm is challenged, it is not difficult to find early observers who tried to reshape its one-dimensional images, seeking like Michelangelo to release the conceptual angel from the surrounding marble. Such an early sculptor was James B. Conant, who as Harvard's president served as one of Bush's closest colleagues during the war. Conant declined to be named the founding director of the National Science Foundation when the new agency was created in 1950. But he agreed to join the National Science Board and was elected its chairman. His foreword to the Foundation's first annual report included this notably heterodox view:

> No one can draw a sharp line between basic and applied research and the Foundation will support many investigations that might be classed in one area or the other. Indeed, speaking for myself and not for the Board, I venture to suggest that we might do well to discard altogether the phrases "applied research" and "fundamental research." In their place I should put the words "programmatic research" and "uncommitted research," for there is a fairly clear distinction between a research program aimed at a specific goal and an uncommitted exploration of a wide area of man's ignorance. It would be safe to say that all so-called applied research is programmatic *but so, too, is much that is often labeled fundamental.*[2]

Conant made clear that this view was his own and not the board's. Well he might, since the Foundation's annual reports stressed the importance of the "technological sequence." Conant avoided a direct clash with Bush by substituting "fundamental" for "basic." But Conant understood these terms to refer, interchangeably, to all research that seeks to extend understanding within a scientific field—and therefore to include more than the curiosity-driven science that he called "uncommitted research" and Bush had called basic research.[3] Indeed, by refusing to equate "fundamental" with "uncommitted" Conant recognized a cross-cutting relationship between the goals of understanding and use, one that divides basic or fundamental research into *programmatic* work that is influenced by considerations of use and *uncommitted* work that is a pure voyage of discovery.

The idea of dividing basic research according to whether or not it also is inspired by considerations of use has appealed to a number of observers who wanted to provide for a more complex relationship between these goals. The historian of science, Gerald Holton, in his remarkable essay on Thomas Jefferson's vision of the Lewis and Clark expedition, articulates the need for a category of research that combines Newton's tradition of understanding the natural world with Bacon's tradition of using this understanding to achieve purposive ends. Such a category would encompass "research in an area of basic scientific ignorance that lies at the heart of a social problem."[4] Lillian Hoddeson, in a series of articles on basic research in Bell Laboratories, offered this modification of the framework:

> "Fundamental" and "pure research" refer to the attempt by experimental and theoretical means to understand the physical underpinnings of phenomena. The special term "basic research" refers here to fundamental studies carried out in the context of industry, which may lead to, but do not aim primarily at, application. Applied research, on the other hand, which encompasses engineering and technology, does aim primarily at practical application.[5]

Hoddeson's specialized use of "basic research" is close to the category of research offered by Deborah Shapley and Rustum Roy in their dispirited survey of contemporary science and science policy:

> What was lost, in a word, was the importance of applied science and engineering, and something else we shall call *purposive basic research*, i.e., research of a fundamental nature that is done with a general application in mind, like Charles H.

Townes' discovery of the maser while working on microwave transmission for Bell Laboratories, or most biomedical research.[6]

Frustration with the prevailing framework is indeed endemic among those who have tried to fit its categories to research in biomedical science. A number of biomedical scientists have argued that applied research includes studies that also seek a more basic understanding of a field. Thus Julius Comroe and Robert Dripps, in their seminal study of work leading to major clinical advances, define a category of research that is related to a clinical problem but is also "concerned with basic biological, chemical, or physical mechanisms."[7]

The insulation of basic research from thought of practical ends has been defended against such challenges partly by conceding the legitimacy of the concern for applied goals among those who *support* research but not among those who *perform* it. In an era of institutionalized science, research is typically set in an organizational framework where influence on goals may be shared with those who establish priorities and control funds at various levels. Alan T. Waterman, NSF's first director, wove this difference between sponsor and investigator into a defense of Bush's belief that scientists must be free to pursue basic research wherever it leads. In his 1964 address as retiring president of the American Association for the Advancement of Science, Waterman noted:

There has been a steady increase in the support of basic research which may be termed "mission-oriented"—that is, which is aimed at helping to solve some practical problem. Such research is distinguished from applied research in that the investigator is not asked or expected to look for a finding of practical importance; he is still exploring the unknown by any route he may choose. But it differs from "free" basic research in that the supporting agency does have the motive of utility, in the hope that the results will further the agency's practical mission ... Thus, basic research activity may be subdivided into "free" research undertaken solely for its scientific promise, and "mission-related" basic research supported primarily because its results are expected to have immediate and foreseen practical usefulness.[8]

It is noteworthy how deftly Waterman introduced the category of "mission-oriented" basic research without giving an inch on Bush's insistence that basic research must be done by scientists who have no thought of practical ends. In Waterman's formulation, only the funding agency need have such thoughts, as it supported "mission-oriented" basic research. The individual investigator would, in effect, share with the sponsoring agency only the choice of the research problem, and thereafter be free to pursue the research without thought of practical ends.

Harvey Brooks offered a more sophisticated version of Waterman's view in a 1967 report to the House Committee on Science and Astronautics on how to enlist science for advances in technology as Congress took up the "Daddario amendments" to NSF's charter.[9] Brooks's introduction sets out an interesting analysis of the distinctness of basic and applied research, one that echoes Waterman by noting that

> there can be a perfectly viable difference in viewpoint between the research worker and his sponsor. Research that may be viewed as quite fundamental by the performing scientist may be seen as definitely applied and may fit into a coherent pattern of related work from the standpoint of the sponsoring organization or agency.[10]

This observation led Brooks, as it had Waterman, to subdivide basic research according to this interplay of institutional influences on problem selection. Shortening Waterman's "mission-oriented basic research" to "oriented basic research," he observed that

> the general field in which a scientist chooses or is assigned to work may be influenced by possible or probable applicability, even though the detailed choices of direction may be governed wholly by internal scientific criteria. Research of this type is sometimes referred to as "oriented basic research."[11]

Brooks also noted that research may be differently perceived according to where it is done. For example, certain types of research on semiconducting materials, carried out in a university laboratory, "might be regarded as fairly 'pure,' while in Bell Laboratories they would be regarded as 'applied' simply because potential customers for the research results existed in the immediate environment,"[12] a factor that influences the view held by the bench scientist and not only the view held by the scientist's sponsors:

> Once the transistor was discovered, and germanium became technologically important, almost any research on the properties of group IV semiconducting materials could be considered to be potentially applicable . . . and research into the theory of zone-refining single crystals was of such obvious immediate application to the control of transistor materials that it could legitimately be called applied rather than merely applicable," whereas "prior to the discovery of the transistor, both of these types of research would have been of equal interest and importance from the scientific viewpoint, but they would have been classified as quite fundamental or 'pure'.[13]

But in a remarkable aside, Brooks allowed himself a far more radical view by noting that

> the terms *basic* and *applied* are, in another sense, not opposites. Work directed toward applied goals can be highly fundamental in character in that it has an important impact on the conceptual structure or outlook of a field. Moreover, the fact that research is of such a nature that it can be applied does not mean that it is not also basic.[14]

He supported this observation with the example of Louis Pasteur, whose later work was, as we have seen, an impressive synthesis of the goals of understanding and use. This aside represented a much more radical break with the idea of a one-dimensional spectrum of basic and applied research and helps to prepare the way for a different framework for thinking about the goals of understanding and use.

Official Reporting Categories

With the interests served by Bush's framework so firmly entrenched, the United States did little at an official level to respond to the logic of these early dissents. But the countries with a different postwar experience sought to recognize a more complex relationship between understanding and use. With the exception of Britain, none of the other industrial countries shared the distinctive circumstances that led the postwar paradigm to become so deeply ingrained in America.[15] The economic and social dislocations of the war kept the scientists in these countries from making claims on government equivalent to those asserted in the United States by the campaign that followed publication of *Science, the Endless Frontier*, although the postwar stature of American science made Bush's framework highly visible in all of the industrial countries.

A natural, if ultimately limited, focus for the conceptual efforts to mix the goals of understanding and use was the work of the Organization for Economic Cooperation and Development to refine the categories within which OECD's member nations reported scientific and technological activities. These efforts can be traced through the successive versions of OECD's *Frascati Manual*, so called because the 1963 conference that agreed on the first manual was held in the Italian town of Frascati. The first manual, drafted largely by Christopher Freeman, a British specialist on science policy who later was cofounder of the Science Policy Research Unit at the University of Sussex, drew on the definitions the U.S. National Science Foundation had been using for about a decade. Hence, his draft presented no challenge to the Bush categories, to the relief of the National Science Foundation's representatives. *Fundamental research* was defined as "work undertaken primarily for the advancement of scientific knowledge, without a specific practical application in view;" *applied research*, as work that did have "an application in view." Moreover, in keeping with the linear model of technology transfer, *experimental development* was defined as "the use of the results of fundamental and applied research directed to the introduction of useful materials, devices, products, systems, and processes, or to the improvement of existing ones;"[16] Bush's second canon, that technological innovation is ultimately rooted in scientific discovery, was alive and well at the Frascati conference.

These categories were modified when the *Frascati Manual* was revised in 1970. This revision approached the definition of basic and applied research at three levels. It first of all offered a generic definition of *research and experimental development* as "creative work undertaken on a systematic basis to increase the stock of scientific and technical knowledge and to use this stock of knowledge to devise new applications."[17] It then defined *basic research* ("fundamental" having given way to Bush's term) as "original investigation undertaken in order to gain new scientific knowledge and understanding . . . not primarily directed towards any specific practical aim or application" and *applied research* as "original investigation undertaken in order to gain new scientific or technical knowledge . . . directed primarily towards a specific practical aim or objective."[18] Thus far, the prevailing framework remained unchallenged.

At a third level, however, the revised manual added some observations about basic research that echo Waterman's and Brooks's view of "oriented research," noting in particular that, although basic research "has no immediate specific practical applications in view," it "may be oriented towards an area of interest to the performing organization," adding that "in *oriented basic research* the organization employing the investigator will normally direct his work towards a field of present or potential scientific, economic or societal interest."[19] These revisionist comments were accompanied by a figure, reproduced here as figure 3-1, in which a circle of "oriented basic research" is included in a larger circle for "applied research"—as well as, puzzlingly, in a still more embracing circle for "experimental development"—with a circle for "pure basic research" tangent to, but not intersecting, these nested circles.

Although this *fleur-de-lis*-like figure did signal some relaxation in the presumption that understanding and use are opposed, it did little to clarify the conceptual relationship between these goals and has not been widely reproduced. It also suffered the disabilities of providing only for a dichotomous split of basic research between "pure" and "oriented" research and of presuming that the mix of goals in the latter category results only from the organizational

Figure 3-1. *Diagrammatic Presentation of the Concepts of Basic and Applied Research and Experimental Development Research in the 1970 Frascati Manual*

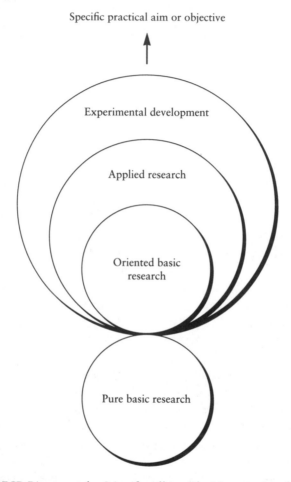

Specific practical aim or objective

Experimental development

Applied research

Oriented basic research

Pure basic research

SOURCE: OECD Directorate for Scientific Affairs, *The Measurement of Scientific and Technical Activities: Proposed Standard Practice for Surveys of Research and Experimental Development (Frascati Manual)* (Paris: Organization for Economic Cooperation and Development, 1970), p. 14.

sponsorship of research and not from a meld of goals held by the research scientist. This organizational gloss is missing from a number of subsequent definitions of "strategic research," the term that supplanted "oriented research" in the 1980 revision of the *Frascati Manual*.[20]

Two British scholars, John Irvine and Ben R. Martin, have addressed the issue of strategic research in the course of two illuminating surveys of research foresight in a number of countries.[21] Their 1989 book has this to say:

> Here, the traditional three-fold distinction between "basic research," "applied research," and "experimental development" is now recognized as inadequate. The "basic" category is especially problematic in that it covers a disparate variety of activities ranging from curiosity-oriented, proposal-driven research through long-term targeted programmes supported by sectoral government agencies, to speculative work in industry where no specific application is yet in mind. It is therefore useful to subdivide "basic research" into "curiosity-oriented research" and "strategic research."[22]

It has been inherently difficult for governments to resolve the conceptual issue surrounding the goals of research by adumbrating a set of statistical reporting categories. Almost any useful statistical series becomes the prisoner of its existing definitions, and the difficulty of establishing the motives of scientific research has strengthened the hand of those who have wanted to preserve the empirical separateness of basic and applied research. Hence, the conceptual issue of strategic research has been taken hostage by problems of measurement and has remained unresolved.

This was decidedly so when the U.S. National Science Foundation considered the possibility of revising the Bush framework. The backdrop to this episode was the willingness of Congress and the Reagan administration to establish new programs of Engineering Research Centers and Science and Technology Centers. These centers were typically located in universities but with the participation of industry and the state governments and were designed to bring the resources of several scientific and engineering disciplines to bear on problem areas of evident importance for the country's needs.

It is hardly surprising that as the centers took root, NSF's director, Erich Bloch, should wonder whether the categories for reporting government-funded R&D adequately provided for strategic research of the kind the centers were intended to mount.

Bloch therefore created a task force to consider this issue and requested that the group also examine the mixed taxonomies proposed by the British government and U.S. General Accounting Office. The task force's report clearly signaled the importance of the newly funded centers in launching this review:

> In recent years a number of new "research centers" have been formed, often as a partnership of Federal government, state government, industrial and academic interests. The research performed at these centers tends to combine many traditional disciplines and is oriented toward generating knowledge in fields that may lead to discoveries that will enhance the strategic position of the U.S. in the world economy. . . . The existing taxonomy of research does not address this type of research very well.[23]

The task force did not address the conceptual issue head-on but shifted the basis of a taxonomy from the goals of research to the intended *users* of research.[24] It proposed a threefold categorization of research—with *fundamental research* leading to "results intended at the time the research is funded for dissemination to other researchers and educators"; *strategic research* to results "of evident interest to a broad class of users, external to the research community, that can be identified at the time the research is funded," although "the intended users of the results may also be within the research community"; and *directed research* to results bearing on "the specific needs of the sponsoring organization."[25]

The report and appended evaluation of other taxonomies made clear that the task force had chosen to focus on users in the belief that it would be difficult to match accurate data to a taxonomy based on goals and that it should find a taxonomy that would be a "nonthreatening" change for other federal agencies that fund substantial R&D; the conceptual issue was again taken hostage by problems of measurement. In any event, not much came of the task force's proposals. NSF still adheres to definitions of basic and applied research that are firmly in the Bush tradition. Only in its annual survey of *industrial* R&D does it attach to the definition of basic research the limited observation that basic research "may be in the fields of present or potential interest to the reporting company."[26]

Hence, the twenty-year effort of the OECD countries to modify their reporting categories has done less than we might expect to clarify the relationship of understanding and use as goals of research. The extensive discussions of a new (fifth) edition of the *Frascati Manual* that were held in the early 1990s found only two governments pressing to modify the traditional distinction between basic and applied by including a category for strategic research. These two governments did not agree on how such a category should be defined, and the reservations among the other members were strong enough to limit the headway that could be made toward resolving the conceptual issue surrounding this distinction. Indeed, language that went beyond the prior *Frascati Manuals* was considerably watered down during the stage of consultation with member countries on the text of the new edition.

The reservations were of several kinds. To begin with, there again was a desire to preserve the historical distinction between basic and applied and the statistical series associated with it. As a result, revisionist proposals were directed toward how strategic research might be accommodated by drawing distinctions *within* the basic and applied categories, rather than by cutting across these categories. There was also the semantic concern that "strategic research" might be confused with national or international security studies, or with research on strategic materials or technologies. But there was at least a faint new concern—that by reporting commercially relevant strategic research an OECD country might be seen by other governments as indirectly subsidizing goods exported by firms that benefited from the results of such research. Some of OECD's members were reluctant to seed a new set of trade disputes by creating a category for reporting strategic research.

With this last concern, the wheel came full circle. The belief that science could be enlisted in the drive toward economic competitiveness had fueled much of the interest in strategic research in OECD's member countries in the first place—and therefore had also fueled much of the interest in defining categories for reporting such research. But the very awareness that strategic research might improve a country's trading position, and therefore be regarded as an export subsidy, ultimately helped to close off the effort to define one or more categories for reporting strategic research. After defining language had been excised from the draft, all that remained in the new Frascati edition was the observation that distinguishing oriented from pure basic research "may provide some assistance towards the identification of strategic research"[27] and the observation that

> while it is recognized that an element of applied research can be described as strategic research, the lack of an agreed approach to its separate identification in Member countries prevents a recommendation at this stage.[28]

If OECD is to play a significant role in clarifying the conceptual issue of the relationship of understanding and use as goals of research, it awaits a new *Frascati Manual* in 2000 before some fresh sculpting of categories can free this conceptual angel from the statistical marble. What is needed is a way of cutting through the inherently ambiguous choice of assimilating strategic research *either* with basic *or* with applied research. Let us see how this problem can be resolved by a framework that is clear and conceptually spare.

Expanding the Dimensional Image

So strong is the hold of the one-dimensional basic-applied spectrum that many observers who find it difficult to fit this framework to the realities of research think the problem must be because of the uncertainty of classification near the middle of such a spectrum, as if they were measurement psychologists seeking to discriminate two latent classes of subjects on the basis of unreliable measurements on a single scale. In this vein, a former director of the Division of Science Resource Studies of the National Science Foundation has said of the basic-applied spectrum that

> any process that divides a continuum into discretely demarcable regions is generally plagued by fuzziness and overlaps at the boundaries of the subdomains.[29]

Figure 3-2. *One Hypothetical Placement of Pasteur on the One-Dimensional Basic-Applied Spectrum*

Basic   0   Applied

But the difficulty here is more than "fuzziness and overlap at the boundaries." It lies rather in the attempt to force into a one-dimensional framework a conceptual problem that is inherently of higher dimension.

To trace the implications of this we may note that Conant and other physical scientists who have wanted to divide basic research according to whether it also is guided by applied ends have implicitly seen a cross-cutting relationship between the goals of understanding and use. And Comroe and Dripps and many of the other life scientists who have wanted to divide applied research according to whether it also seeks a more fundamental understanding have likewise seen a cross-cutting relationship between these research goals.

To see how this reformulation would go, return to the familiar idea of a spectrum of research that extends from basic to applied and ask *where* on this spectrum should one place the mature Pasteur? The first instinct might be to place him at the mid- or zero-point of the spectrum in view of his commitment to both understanding and use (figure 3-2).

But a moment's reflection is enough to see that this is quite wrong and that the mature Pasteur deserves to be placed not at one point but at two: he belongs far to the left of the spectrum in terms of the strength of his commitment to *understand* the microbiological processes he discovered, but he equally belongs far to

Figure 3-3. *A Second Hypothetical Placement of Pasteur on the One-Dimensional Basic-Applied Spectrum*

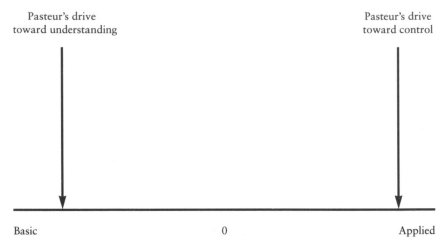

the right of the spectrum in terms of the strength of his commitment to *control* the effects of these processes on various products and on animals and humans (figure 3-3).

We have therefore the anomaly of Pasteur's being represented by two Cartesian points in this Euclidean one-space, an anomaly that should lead us to wonder whether such a one-dimensional figure can adequately characterize research in terms of its basic and applied goals. We may remove this anomaly while still retaining the ease of interpreting a space of spare dimension if we grasp the spectrum at its zero point, rotate the left-hand half through an arc of 90 degrees, and restore Pasteur to the status of a single Cartesian point in what is now a two-dimensional conceptual *plane* (figure 3-4). The vertical axis represents the degree to which a given body of research seeks to extend the frontiers of fundamental understanding, the horizontal axis the degree to which the research is guided by considerations of use.

There is not the slightest reason to think of these dimensions only in dichotomous terms, since there can be many degrees of commitment to these two goals. But if we do so for heuristic reasons, it is clear that we now have not one dichotomy but two. This dual dichotomy can be exhibited as a fourfold table with cells or quadrants (figure 3-5).[30]

Figure 3-4. *Pasteur's Placement in a Two-Dimensional Conceptual Plane*

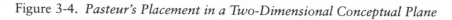

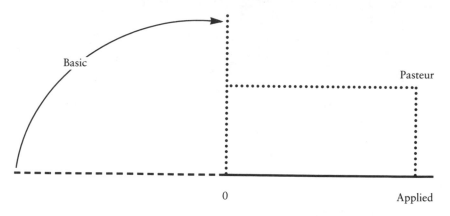

It will help to fix the meaning of this array if we characterize its quadrants. The upper left-hand cell includes basic research that is guided solely by the quest for understanding without thought of practical use. It might be called Bohr's quadrant in view of how clearly Niels Bohr's quest of a model atomic structure was a pure voyage of discovery, however much his ideas later remade the world. This category represents the research ideal of the natural philosophers, institutionalized in the pure science of the Germans

Figure 3-5. *Quadrant Model of Scientific Research*

Research is inspired by:

		Considerations of use?	
		No	Yes
Quest for fundamental understanding?	Yes	Pure basic research (Bohr)	Use-inspired basic research (Pasteur)
	No		Pure applied research (Edison)

in the nineteenth century and of the Americans in the twentieth, and includes Bush's concept of "basic research."

The lower right-hand cell includes research that is guided solely by applied goals without seeking a more general understanding of the phenomena of a scientific field. It would be appropriate to call it Edison's quadrant, in view of how strictly this brilliant inventor kept his co-workers at Menlo Park, in the first industrial research laboratory in America, from pursuing the deeper scientific implications of what they were discovering in their headlong rush toward commercially profitable electric lighting. A great deal of modern research that belongs in this category is extremely sophisticated, although narrowly targeted on immediate applied goals.

The upper right-hand cell includes basic research that seeks to extend the frontiers of understanding but is also inspired by considerations of use. It deserves to be known as Pasteur's quadrant in view of how clearly Pasteur's drive toward understanding and use illustrates this combination of goals. Wholly outside the conceptual framework of the Bush report, this category includes the major work of John Maynard Keynes, the fundamental research of the Manhattan Project, and Irving Langmuir's surface physics. It plainly also includes the "strategic research" that has waited for such a framework to provide it with a conceptual home, a case of orphanhood noted above.

The lower left-hand quadrant, which includes research that is inspired neither by the goal of understanding nor by the goal of use, is *not* empty, and the fact that it is not helps make the point that we do have two conceptual dimensions and not simply a more elegant version of the traditional basic-applied spectrum. Indeed, the "prediction" of such a category further validates the framework as a whole. This quadrant includes research that systematically explores *particular* phenomena without having in view either general explanatory objectives or any applied use to which the results will be put, a conception more at home with the broader German idea of *Wissenschaft* than it is with French and Anglo-American ideas of science. Research of this type may be driven by the curiosity of the investigator about particular things, just as research in Bohr's quadrant is driven by the curiosity of the scientist about more general things. The bird watchers who are grateful for the highly systematic research on the markings and incidence of species that went into *Peterson's Guide to the Birds of North America* might want to call this Peterson's quadrant, although this is too limited an example to warrant the name.

In the dynamic pathways that link research in the four cells of the table, it is clear that studies in the fourth quadrant can be important precursors of research in Bohr's quadrant, as it was in the case of Charles Darwin's masterpiece *The Origin of Species*, as well as of research in Edison's quadrant. Other motives inspire research in this quadrant. There are cases in which the prime goal of research is to enhance the skills of the researchers. Arnon gives examples of agricultural research projects in which the investigators start to work in a new area not for the findings they will obtain but to gain skill and experience they may later use "when problems arise in the area" or when breakthroughs achieved by other researchers make the field hot.[31] Those familiar with the role of research in the policy process will have no difficulty identifying cases where studies are launched not for what they learn but to block the start of an operating program, a goal to which the investigators may be willing parties.[32]

Probing the Framework

The sense of abstractness is lessened and the greater realism of such a conceptual plane is demonstrated if this framework is applied to an illustrative body of research. A chapter from the annals of research that admirably lends itself to this purpose is the analysis by Comroe and Dripps of the developments in physical and biological science that led to the most significant recent advances in diagnosing, preventing, and curing cardiovascular or pulmonary disease.[33] These investigators mounted their uniquely detailed inquiry into the scientific backdrop of new technology in the 1970s, provoked by the shift toward purely applied biomedical research that had been signaled by the Johnson and Nixon administrations.

The findings from their meticulous study are, to begin with, a striking illustration of how multiple, unevenly paced, and nonlinear are the paths between scientific discovery and new technology. From this standpoint, their account of the developments leading to cardiac surgery is especially interesting:

When general anesthesia was first put to use in 1846, the practice of surgery exploded in many directions, except for thoracic surgery. Cardiac surgery did not take off until almost 100 years later, and John Gibbons did not perform the first successful operation on an open heart with complete cardiopulmonary bypass apparatus until 108 years after the first use of ether anesthesia. What held back cardiac surgery? What had to be known before a surgeon could predictably and successfully repair cardiac defects? First of all, the surgeon required precise preoperative diagnosis in every patient whose heart needed repair. That required selective angiocardiography, which, in turn, required the earlier discovery of cardiac catheterization, which required the still earlier discovery of X-rays. But the surgeon also needed an artificial heart-lung apparatus (pump-oxygenator) to take over the function of the patient's heart and lungs while he stopped the patient's heart in order to open and repair it. For pumps, this required a design that would not damage blood; for oxygenators, this required basic knowledge of the exchange of O_2 and CO_2 between gas and blood. However, even a perfect pump-oxygenator would be useless if the blood is clotted. Thus the cardiac surgeon had to await the discovery and purification of a potent, nontoxic anticoagulant—heparin.[34]

The aspect of their analysis that directly bears on our framework is their painstaking assessment of the goals moving those responsible for the scientific advances that prepared the way for these breakthroughs in medical technology. Comroe and Dripps first of all elicited from physicians and specialists in the field the ten most important clinical advances since the early 1940s for "diagnosing, preventing, or curing cardiovascular or pulmonary disease; stopping its progression, decreasing suffering, or prolonging useful life." In addition to open heart surgery, the resulting list included blood vessel surgery, treatment of hypertension, management of coronary artery disease, prevention of poliomyelitis, chemotherapy of tuberculosis and acute rheumatic fever, cardiac resuscitation and cardiac pacemakers, oral diuretics (for treatment of high blood pressure or of congestive heart failure), intensive care units, and new diagnostic methods.

They then analyzed the work that led to each of these advances. With the help of 140 consultants they identified the knowledge essential to each advance and more than 500 "key articles," going back in some cases more than two centuries, reporting the work that developed this knowledge. They made these articles (or, equivalently, the reported research) the central focus of their analysis, classifying them in two ways. The first was whether the authors of these reports gave any sign of the work's having been "clinically oriented" by indicating "an interest in diagnosis, treatment, or prevention of a clinical disorder or in explaining the basic mechanisms of a sign or symptom of the disease itself." In our framework, this amounts to asking whether a given piece of work should be placed in the left- or right-hand *column* of figure 3-5.

Comroe and Dripps crossed this with a second classification, according to whether the reported research was basic in the sense that the investigator sought to understand the mechanisms responsible for observed effects; that is, in our terms, whether the research should be placed in the upper or lower *row* of our fourfold table. The three resulting categories—basic research unrelated to the solution of a clinical problem, basic research related to the solution of a clinical problem, and research not concerned with basic biological, chemical, or physical mechanisms—correspond with Bohr's, Pasteur's, and Edison's quadrants. They found that these categories included, respectively, 37 percent, 25 percent, and 21 percent of the key articles. The remaining 17 percent were classified as development (15 percent) or as "review and synthesis" (2 percent). The 25 percent classified as basic research related to the solution of a clinical problem (that is, work lying in Pasteur's quadrant) is impressive further evidence of the intermingling of understanding and use as goals of research, although the 37 percent classified as basic research unrelated to the solution of a clinical problem (that is, work lying in Bohr's quadrant) is a fresh tribute to the role of pure research in new technology.[35]

What is entailed by this way of thinking about basic and applied science may be further clarified by addressing four conceptual issues. Each is important in its own right, and a discussion of these points may also lessen any sense that invoking the idea of a conceptual plane is a purely formal device.

Characterizing research ex ante or ex post. The first of these issues is whether the classification of research as basic and applied should rest on advance judgments as to the intended goals of research or on retrospective judgments as to what research has achieved. It is sometimes objected that classifying research on the basis of intended goals involves unscientific speculation about the motives of researchers that is quite unlike the assured and objective judgments historians of science can later make. One resisting scientist has said that distinguishing basic from applied research on the basis of such ex ante judgments is like putting scientists on the couch.

The logic of classifying research on the basis of intended goals rather than known achievements rests on the fact that policy has to do with choice—the choices facing individual scientists, the choices facing those who match resources to alternative research uses at the retail or wholesale level. All of these require ex ante judgments under the uncertainty that is an inherent part of research yet to be done. Although the historian of science will in due course be able to give far more assured judgments as to which research proved in fact to advance the general understanding of a field and which in fact led to significant use, only a framework that deals ex ante with the goals of research can serve the needs of science and technology policy.

Such an approach reaches beyond purely private motives. Although there must always be some uncertainty as to whether the goals of research will be achieved, these purposes have to do with "objective" future conditions, about which considered judgments can be made. Indeed, the integrity of the peer review process rests on the fact that it is possible to reach considered, institutionalized judgments on the likelihood of achieving the goals specified for particular projects of research.

Whose goals are to be consulted? Sometimes the objection is made that it is impossible to distinguish types of research on the basis of goals because those who play different roles in the modern system of research may have different goals for a given project. In an era of organized science, research is, as already noted, typically done in an institutional framework where influence on goals may be shared with those who set priorities and control funds at various organizational levels. The sharper focus of working scientists on understanding and of their sponsors on use is a conspicuous element of a system that involves heavy government support. A university-based biomedical scientist seeking support from the National Institutes of Health may see the proposed work as extending fundamental knowledge, and so may the investigator's department head and the NIH study section that recommends support. Yet the project may be approved by the university's vice president for medical affairs and funded by NIH, and ultimately by Congress, for the contribution it makes to the control of disease. Some years ago Charles V. Kidd wryly noted that the nation's universities reported accepting $85 million in federal grants for basic research in a year in which the government thought its outlays for basic research were half that large.[36] Vannevar Bush, after all, recognized the difference of view between scientist and sponsor on the grand scale when he called on the nation to advance its social and economic goals by supporting research that would, in an immediate sense, be driven only by the scientist's quest of added understanding. As noted earlier, Alan T. Waterman, NSF's first director, wove this difference of view between investigator and sponsor into a defense of Bush's belief that scientists must be free to pursue basic research wherever it leads. In one sort of limiting case, a sponsor might put together a portfolio of basic studies involving multiple researchers without letting the researchers in on the applied objective.

Yet the point should be forcefully made that the mix of goals in use-inspired basic research is not *only* the result of differing goals being held by those at different levels of the institutionalized system of modern science. Despite the rearguard action by Waterman and others to defend the purity of the quest of understanding by the individual scientist, the annals of research are replete with examples of work by investigators who were directly influenced both by the quest of general understanding and by considerations of use. Pasteur wanted to understand *and to control* the microbiological processes he discovered. Keynes wanted to understand *and to improve* the workings of modern economies. The physicists of the Manhattan Project wanted to understand *and to harness* nuclear fission. Langmuir wanted to understand *and to exploit* the surface physics of electronic components. The molecular biologists have wanted to understand *and to alter* the genetic codes in DNA material.

Moreover, the sharing of influence on research choices between working scientist and sponsor need not entail so sharp a dissonance on goals as to make it unreasonable to classify research within our two-dimensional framework. In the major scientific countries the independence of university-based scientists is well enough established that they largely set their own goals within inevitable resource constraints and the perspectives of their scientific disciplines, which typically dominate the peer-review mechanisms for allocating grants. Yet this independence has not precluded a lively interest in applied goals in academic fields as diverse as chemistry, computer science, economics, molecular biology, pharmacology, statistics, and atomic, molecular, and optical science. If academic scientists have a deserved reputation for pursuing interests of their own, they too are generally faithful to added objectives when they become involved in basic, use-inspired sponsored research. Likewise, scientists who work in government or industrial laboratories generally accept the mission of these units, even if many retain a taste for basic science—one that is encouraged by the leadership of the strongest of these laboratories as a means of recruiting, developing, and retaining excellent staff.

The substantial volume of basic academic research that is use inspired helps to explain the ironic inequality noted by Kidd—that the universities reported accepting a total of federal grants for basic research twice as large as the government thought it made in a particular year; since the accounting both by the universities and by the government used an either-or coding of basic and applied research, it should not be surprising that the universities considered much of their federally funded Pasteur's quadrant research to be basic, while the government considered much of it to be applied.[37] The dissonance as to goals between working scientists and their overseers or funders would be diminished if it were generally perceived that research can be simultaneously influenced by the quest of scientific understanding and considerations of use, a point that deserves special emphasis:

Freed from the false, "either-or" logic of the traditional basic/applied distinction, individual scientists would more generally see that applied goals are not inherently at war with scientific creativity and rigor, and their overseers and funders would more generally see that the thrust toward basic understanding is not inherently at war with considerations of use.

Indeed, the institutional settings of modern science do not produce conflict over research goals so much as help to define these goals for their scientific staff. This conclusion reverses the thrust of the reservation as to whose goals should be consulted. The organizational settings of research do not so much complicate a goal-based framework for thinking about science and technology policy as they encourage research with particular patterns of the alternative goals of understanding and use, including research that is both basic and use-inspired. For example, a number of research units, some within industry (Bell Laboratories), some free-standing (the Rand Corporation), some within the universities, have used a matrix plan of organization to engage first-class scientists in research of impressive scientific rigor that is also deeply influenced by considerations of use.

Can the two dimensions be reduced to one? The graphic image of a one-dimensional basic-applied spectrum naturally gives way to the two-dimensional plane once it is clear that these goals are not inherently opposed. But the power of one-dimensional thinking is such that there have been other attempts to array research on a single scale. An instructive effort of this kind was included in a 1981 report of the Australian Science and Technology Council (ASTEC).[38] The report reproduces the categories of research proposed by the *Frascati Manual*, with modifications by Australia's Bureau of Statistics. These definitions, as already noted, move toward a cross-cutting vision of basic and applied research. But the report fails to pursue this logic and instead proposes a single-dimensional research spectrum that extends from "immediately applicable" to "highly abstract." The graphic representation of this spectrum is reproduced here as figure 3-6. The relative locations of pure, strategic, and tactical research are suggested by three Gaussian (bell) distributions that march across this Euclidian one-

Figure 3-6. *Australian Modification of Linear Model*

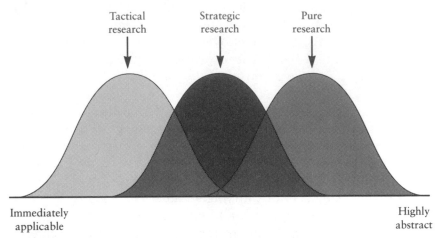

SOURCE: Australian Science and Technology Council, *Basic Research and National Objectives* (Canberra: Australian Government Publishing Service, 1981), p. 6. Commonwealth of Australia copyright, reproduced by permission.

space from the "immediately applicable" to the "highly abstract" poles.

This spatial imagery owes more to the appeal of one-dimensional thinking than to the characteristics of research that the ASTEC authors sought to bring out. By labeling one of their poles "immediately applicable," the drafters show their desire to contrast the two goals of research described by their modified Frascati definitions. But it simply clouds the issue to make "highly abstract" the pole opposite from "immediately applicable" and to produce another one-dimensional array. Abstract thinking is no doubt most conspicuous in research that lies in Bohr's quadrant and least prominent in work that lies in Edison's quadrant. But this is only a statement about an empirical correlate of the goal patterns envisaged by the Frascati definitions. The ASTEC authors would have been closer to the mark if they had abandoned their one-dimensional framework and used their graphic skills to illuminate the conceptual basis of the distinctions among "pure," "strategic," and "tactical" research.

Time to application. The most important factor that is some-times believed to array research on a single continuum is the idea of "time to application." Indeed, this factor is often thought to define the difference between basic and applied research. No one can doubt that there is a vast difference in the time that is likely to elapse between the production of new knowledge and the uti-lization of this knowledge for an applied purpose as one moves from Bohr's to Edison's quadrants. It could hardly be otherwise, since pure basic research seeks only to probe unknown fundamen-tals, while purely applied research seeks only to meet some clearly defined need. But time to application is far more problematic in the use-inspired basic research of Pasteur's quadrant, which seeks both to probe unknown fundamentals and to meet a societal need. The knowledge gained by Pasteur's own fundamental work in microbiology was quickly applied to industrial and public health problems, as much of the fundamental work in molecular biology is quickly applied in biotechnology today—indeed, so quickly that some observers speak playfully of negative time to application. However, the plasma scientists will in the end require more than half a century to gain the basic understanding that will yield com-mercially profitable power from nuclear fusion. Moreover, there is a good deal of variation not only in time-to-application but in our ability to estimate the time horizon of application. It may indeed make sense to regard time-to-application and the predict-ability of this time as separate dimensions. The reasonable view is therefore that time to application is not a one-dimensional sub-stitute for our conceptual plane but an important empirical cor-relate of the pattern of goals that defines this two-dimensional framework.

It will be important to have a clear view of the relationship between time and use to understand the policy implications of this framework, which are discussed in chapters 4 and 5. It will be especially important to see that some advances of fundamental scientific importance have near-term applications—and not to think that all research of a basic character must play only a distant role in advances in technology. This point is far more easily grasped if one appreciates the reality of use-inspired basic research, a reality expressed by Pasteur's quadrant. If we are aware of how often considerations of use, including the needs of evolving technology, do influence fundamental research, it will be easier to understand that this research can have applications in a relatively near future.

But it will also be important to see that considerations of use may influence basic research that is unlikely to bring an early return in technology—and not to suppose that all research with a distant horizon of use must be curiosity-driven science that lies in Bohr's quadrant. To believe that all research with such a time horizon is a pure venture in understanding, whose applications are impossible to foresee, is again to miss the essential point that, where the applications of fundamental science are concerned, "everything good does not have to start with a twinkle in a basic researcher's unfocused eye."[39]

Rethinking the Dynamic Paradigm

To recognize the possibility of use-inspired basic science is to see the role of science in new technology from a perspective quite different from the postwar paradigm's view of basic research as a remote dynamo of technological innovation. Although a degree of metaphorical license will always be needed to organize our thinking about these complex relationships, it is clear that the license extended to the "linear model" running from basic to applied research and on to development and production and operations has long since expired. In the words of Nathan Rosenberg, "everyone knows that the linear model of innovation is dead,"[40] even if it still lives on in parts of the science and policy communities and broader public. It has been dealt mortal wounds by the spreading realization of how multiple and complex and unequally paced are the pathways from scientific to technological advance; of how often technology is the inspiration of science rather than the other way round; and of how many improvements in technology do not wait upon science at all.

Indeed, the last of these criticisms has led a number of observers to shift their focus away from the links between science and technology as such to *all* of the sources of technological innovation, with only a secondary interest in how many of these are ultimately traceable to science. The rapidly expanding literature of innovation has offered a number of alternative images of these sources that are vastly less simplistic than the linear model. Ryo Hirasawa, for example, has proposed a "concurrent system" model of the overlapping rather than sequential management of the phases of research, development, production, and sales by innovative Japanese firms.[41] Hirotaka Takeuchi and Ikujiro Nonaka use a sports idiom to develop this distinction, contrasting a relay race, in which a baton is passed from one runner to the next at the end of each lap, with a rugby game, in which the outcome depends on a team that "tries to go the distance as a unit, passing the ball back and forth."[42] Stephen J. Kline and Rosenberg offer an iterative "chain-linked" model of innovation that distinguishes chain, feedback, and initiation elements, a model which, if nothing else, conveys the potential complexity of the innovation process.[43]

Inevitably, the more general canvass of the sources of innovation involved in these models has somewhat diverted attention from the relationship between basic science and technological innovation. This relationship was what the linear model was all about, however flawed its account. Attention has also been diverted from this relationship by the concern with economic competitiveness, which has led a number of commentators to shift their focus to the link between new technology and its commercial application. Erich Bloch, former director of the National Science Foundation, and David Cheney, a colleague at the Council on Competitiveness, express this concern well:

> Technology that remains in the lab provides almost no economic benefits. Technology that is applied only to government markets, such as defense, provides much smaller economic benefits than technologies that contribute to success in the much larger commercial markets, and especially in the ever more important global markets.[44]

In the view of these authors, the United States leads the world in basic science and probably also in technological innovation; it is falling down, however, in converting new technology into products and services that meet the test of the market. It is almost a commonplace of commentaries on America's lagging competitiveness how often technologies first developed in the United States have been commercially exploited elsewhere in the world, especially Japan. The authors of a comprehensive report on European policy toward innovation and technology diffusion also distinguish between new technology and its use in products or services that meet the test of the market.[45] The British government has incorporated this distinction into the vocabulary of technology policy, using "innovation" for the development of new technology, "exploitation" for its commercial application.

The case for drawing such a distinction would seem to be strengthened by the notable examples from the annals of technology, detailed by Rosenberg and others, in which it took many years for a new technology to find its most important commercial uses. The steam engine was initially seen as a device for pumping water from mines and only later as a power plant for movable ships or carriages. The railroad was initially seen as a feeder of goods for canal transport and only later as a fully articulated system of transportation in its own right. The radio was initially seen as a "wireless" substitute for the electric telegraph for communicating between two points that could not be connected by wire, such as ship to shore, and only later as a means of "broadcasting" communication to a mass audience.[46] Indeed, this is an almost universal phenomenon in the evolution of technology. New technological paradigms seldom spring full-blown from the minds of their inventors, and when they do, as in the case of Arthur Clarke's vision of communications satellites, the visionary is unlikely to be the person who makes the technological dream come true.

Yet there are pitfalls in distinguishing a technology from its applications. A valid distinction is to be drawn between a general technology and its application to particular products or processes. Moreover, particular goods or services may combine several technologies, and some aspects of marketing and finance that may be critically important for economic success are quite distinct from the technology that is being exploited. But it is simply a holdover of linear-model thinking to suppose that technology is shaped only by technical or engineering considerations, free of market influence. Technology itself can be deeply influenced by consumer demand in emerging markets, as it was in each of the cases of the steam engine, railroad, and radio; the technology of the steam locomotive had moved considerably beyond the steam technology that pumped water from the mines. It makes perfectly good sense to speak of a "trajectory" of technology that is guided by technical and by market considerations, as we might speak of the trajectory followed by a branch of science that is guided by several influences—including, at times, the opportunity to create a commercially successful technology.[47]

Indeed, this useful metaphor may be adapted to restate the dynamic problem in these terms:

> To replace the linear model of the postwar paradigm, we need a clearer understanding of the links between the dual but semiautonomous trajectories of basic scientific understanding and technological know-how.

Although the linear model saw the advances of science as fully determining the development of technology, we have seen that the relationship between the two is a far more interactive one, with technology at times exerting a powerful influence on science. It is here that the problems of transforming the static and dynamic paradigms come together: a deeper understanding of this relationship is possible if the dynamic importance of research in Pasteur's quadrant is noted.

Although it would be playful to see a double helix in the intertwined, upward course of scientific understanding and of technological capacity, the one-dimensional, one-way model of the link between basic science and technological innovation clearly needs to be displaced by an image that conceives of their dual, upward trajectories as interactive but semiautonomous (figure 3-7). These trajectories are only loosely coupled. Science often moves from an existing to a higher level of understanding by pure research in which technological advances play little role. Similarly, technology often moves from an existing to an improved capacity by narrowly targeted research, or by engineering or design changes, or by simple tinkering at the bench, in which fresh advances in science play little role. But each of these trajectories is at times strongly influenced by the other, and this influence can move in either direction, with use-inspired basic research often cast in the linking role. In a similar vein, Brooks has observed that "the relation between sci-

Figure 3-7. *A Revised Dynamic Model*

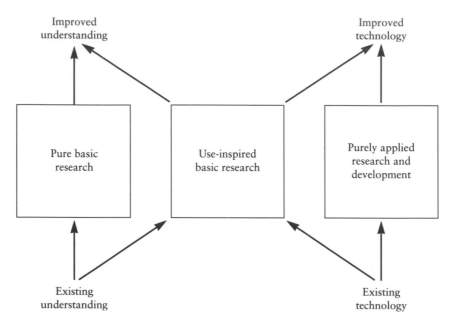

ence and technology is better thought of in terms of two parallel streams of cumulative knowledge, which have many interdependencies and cross-relations, but whose internal connections are much stronger than their cross connections."[48]

This image of dual trajectories of knowledge leaves a good deal out of account. The interaction of science and technology includes the role that new research technologies at times play in the creation of operational technologies and the importance that the availability of commercialized measurement methods may have in supporting new fundamental science. Nonetheless, a loosely interactive relationship has characterized the trajectories of scientific understanding and technological capacity since the period in the nineteenth century when the concept of "technology" first took root in positivist thought. In this period the marriage of science with the practical arts proposed by Francis Bacon more than two centuries before was at last consummated by the influence of technology on the development of science and by the influence of the emerging disciplines of physics, chemistry, and biology on the development of new products and processes.

Implications for Policy

More is involved in these revised images of the links between basic science and technological innovation than their greater faithfulness to the annals of research. These revisions in the postwar paradigm are also of broad importance for science and technology policy. Indeed, the following five observations may carry us across the threshold between analysis and policy:

—The paradigm view of science and technology that emerged from World War II gave a notably incomplete account of the actual relationship between basic research and technological innovation.

—The incompleteness of the postwar paradigm is impairing the dialogue between the scientific and policy communities and impeding the search for a fresh compact between science and government.

—A more realistic view of the relationship of science and technology must allow for the critically important role of use-inspired basic research in linking the semiautonomous trajectories of scientific understanding and technological know-how.

—A clearer understanding by the scientific and policy communities of the role of use-inspired basic research can help renew the compact between science and government, a compact that must also provide support for pure basic research.

—Agendas of use-inspired basic research can be built only by bringing together informed judgments of research promise and societal need.

The policy implications of these observations are discussed in chapters 4 and 5. Chapter 4 explains how a more realistic view of the relationship of basic research to technological innovation can help restore the compact between science and government. Chapter 5 explores how to link judgments of scientific promise with judgments of social value in the funding of basic research that is inspired by considerations of use.

4 | RENEWING THE COMPACT BETWEEN SCIENCE AND GOVERNMENT

T HE DISARRAY OF science and technology policy in the industrial world is especially clear in the United States as the twentieth century draws to a close. It is easy to exaggerate the consensus that underlay the golden age of American science after World War II. The verities that now seem to have commanded universal respect were only tentatively embraced at the time, and the unparalleled postwar flow of funds to basic science moved through organizational channels quite different from those proposed by Vannevar Bush in his epic report, *Science, the Endless Frontier*. Moreover, the Vietnam War, concern for technology's impact on the environment, and the desire to make headway on the nation's unfinished domestic agenda shook the foundations of public support for pure as well as applied science in the late 1960s and early 1970s more than is remembered today; indeed, the Vietnam experience created intense frustration at how little the country's technological edge seemed to weigh on the scales of military success. In view of this ebb in support of science, the Bush era might be said to have lasted twenty-five rather than fifty years.[1]

But the postwar compact made a comeback in the later 1970s

and 1980s, from the time when the Soviet invasion of Afghanistan again focused America's attention on its global adversary. A fresh sense of danger marked the last Carter years and was a prominent element of the early Reagan years. Indeed, the Reagan administration, as much as the Eisenhower administration thirty years earlier, believed that science was an ultimate source of the country's strength as it confronted Soviet power across a still-divided world. As a result, the budgets for basic science made a remarkable recovery in real terms, if they did not soar upward quite as freely as they had in the aftermath of *Sputnik*. Although the postwar compact rested on weakened foundations and the limitations of its underlying paradigm of basic science and technological innovation became increasingly clear, it has crumbled only in the 1990s.

Collapse of the Postwar Bargain

Three developments in the present decade have combined to usher in a profoundly unsettled period in the country's science and technology policy. Each has underscored the limitations in the paradigm underlying the postwar compact between science and government.

End of the Cold War

The first and most conspicuous of the factors in this change was the collapse of Soviet power. The astonishing death of the Soviet empire removed an engine of science policy that started up in two stages at the close of World War II. The final events of the Pacific war led to the view that America's military survival might depend on its leadership in science and technology, and this view was greatly reinforced by the emergence of a global adversary that soon had a bomb of its own. But it was the launching of *Sputnik* that attached a sense of urgency to the need for leadership in technology and in the science believed to underlie it; however well the policy community may have absorbed Bush's dictum in the early years of the Soviet confrontation, the appropriations for basic science remained relatively modest until *Sputnik* created fears that the Soviets might be overtaking American science.

Although these fears receded long before the collapse of the Soviet regime, the Soviet challenge in military technology remained deadly enough into the 1980s for military security still to provide a standing reason for strengthening the country's scientific base. But with the cold war's end, an aura of military survival no longer cloaked the need to assure the scientific source of technological innovation, and this reason for the support of R&D budgets was not reinstated by the subsequent Persian Gulf War, although this conflict created a more favorable impression than Vietnam of the edge that high technology might bring in a local or regional war, one that was encouraged by the Pentagon's successful management of media coverage of the Gulf War.

As this prop of science policy was knocked away, it was quickly evident that a number of leaders of the policy community felt that the investment in basic science should be justified in terms of other societal needs, now that the urgency of military security was greatly reduced, and not simply by Bush's general dictum that advances in fundamental science would nourish the country's future technological advances. Indeed, this was a period in which it was increasingly clear that the bargain between science and government at the end of World War II might prove to be a Faustian one, that a society persuaded to support pure science by the promise of unlimited technological progress might after several decades say, "Now just a moment—we *have* some unmet needs, including some that were *created* by the technological spin-offs of science; the deal is off." There were echoes of such a view in the words of Representative George E. Brown Jr., the chairman of the House Committee on Science, Space, and Technology, as he spoke to his committee:

> From the perspective of policy makers [the perception that the federally funded research system is under] stress may be manifested by discord between the promised benefits of research and a society beset by a range of seemingly intractable economic, environmental, and social problems."[2]

There were echoes in the White Paper on science and technology issued by the British government in the spring of 1993.[3] Others who commanded the congressional budget heights in the first years after the cold war urged the importance of "strategic research" that would more effectively devote the country's research strength to unmet societal needs. In her subcommittee's hearings on the budget for the National Science Foundation, Senator Barbara Mikulski pressed the director to say what fraction of the research funded by NSF was inspired by national needs, and the subcommittee's subsequent report admonished the Foundation to set this fraction at not less than 60 percent.[4]

Integration of the World Economy

Helping America to compete in the global economy was the need most vigorously pressed by the policy community as an appropriate task for the nation's research strength, and the integration of the world economy is a second development that has undermined the postwar compact. By the 1990s the remarkable expansion of world trade under the trading umbrella erected by the United States and its partners after World War II had integrated a great deal of the world's economic activity into a global economy. Despite its economic ascendancy after the war, the United States entered the postwar world with its relatively huge economy remarkably detached from this new trading order; the average of America's imports and exports of goods and services represented only 4.7 percent of gross domestic product even by 1960. This figure had increased more than twofold—to 11.4 percent—by 1994. The comparable figure for the average OECD country grew from 12.5 percent in 1960 to 18.6 percent in 1990.[5] As the global economy became further integrated, the need to compete was given special poignancy for the United States by the progressive loss of manufacturing jobs to other countries. Although most of this loss was to countries with lower labor costs, some was to overseas competitors better able to exploit the new technology, including a number of the science-based technological innovations of American origin. With the cold war winding down, wide portions of the policy community in the United States came to believe that competing in the global economy was the foremost challenge to the country's R&D now that the needs of military security were greatly reduced.

This belief was especially corrosive of the postwar paradigm, because it was so clear that the paramount position of the United States in basic science had not prevented its being severely challenged in world trade. Japan's experience in the postwar decades contradicted Bush's claim that "a nation which depends upon others for its new basic scientific knowledge will be slow in its industrial progress and weak in its competitive position in world trade,"[6] just as Bush's claim was disputed by America's own experience in the earlier decades of the century when it borrowed the European science and technology it needed to become the world's leader in industrial technology while lagging in basic science. Plainly, Japan's success owes more to acquiring and improving upon the world's technology, including a good deal that is science based, than to building a basic-science dynamo to power its industrial progress from within. Some skeptics of a science policy limited to pure research began to paint a dolorous future in which the United States would join the ranks of the world's primary producers, exporting the intellectual ore of its basic research, which other countries would incorporate in high-tech products for sale back to the United States at a price reflecting their added value. Concerns of this sort made a general critique of the prevailing paradigm especially timely, had the intrusion of a third development not slowed the search for a paradigm appropriate for a new science and technology policy.

The Budgetary Legacy of Fiscal and Economic Policy

There is a substantial danger in the current decade that the pressures of the budgetary process will close off efforts to lay the conceptual foundations of a science and technology policy for the coming decades. The intense pressures that will shape the federal budget to the end of the century have built up over a number of years, as a legacy of the fiscal and economic policies pursued by both parties at both ends of Pennsylvania Avenue. Their source lies partly in the belief that the cost of social policy initiatives (and of the Vietnam War) could be met from the dividend provided by the long economic expansion and rising incomes of the 1960s, a miscalculation that was clear when the bills came due in the stagflation of the 1970s; partly in the determination of the Reagan administration to enact the tax cuts of the supply-side economists while building the defense budget in real terms in the 1980s; partly in its failure to persuade Congress to shrink the domestic side of the government and its countenancing of unprecedented interest rates to squeeze inflation from the economy and to attract the foreign funds needed to cover the budget deficits, an experience that converted America from the world's leading creditor to the leading debtor nation and committed a far larger portion of the federal budget to servicing a debt that had tripled in size; and partly in the skill with which later Congresses used the intricacies of the Gramm-Rudman process to put off the hard decisions on budget reduction. Against this backdrop, it was no surprise for the Republicans to put the deficit first when they took control of Congress in 1994. Nor was it a surprise for them to believe they could best achieve drastic reductions by insisting that virtually all discretionary programs share the pain.

In such a climate the opportunities for a more creative science policy are easily missed. Almost inevitably, those making budget decisions that are driven by a broader political agenda have fallen back on established categories and concepts to spell out their detailed policies for the support of research and development by the federal government. Certainly their actions and explanations provide fresh evidence of the hold of the idea of root separation of basic and applied research. Indeed, this distinction has been reinforced by the new congressional leadership's acceptance of the argument from neoclassical economic thinking that pure research should be funded by government as a public good that is unlikely to find alternative support, whereas research and development activities that are nearer to the market should be funded by private firms applying a market test. This outlook put a number of the Democratic administration's technology programs on the block and lessened the share of the pain that the basic research budgets were asked to bear, although very real constant-dollar reductions were promised for these as well. Yet any period of rapid policy change holds the potential for reform, and we should see how a more realistic conception of basic science and technological innovation could help renew the compact between science and government in an era of profound discontinuity in science policy.

The Opening for Renewal

Against the weight of the factors unsettling science and technology policy in the current decade, vigorous public support for basic research is unlikely to be restored by asserting the case for pure science, in accord with the postwar paradigm. Indeed, this assertion increasingly casts the scientific community in the role of an interest group seeking support for an activity that reflects its own essential needs rather than in the role of informed spokesman for an important general interest. However appealing the scientific community may find the ideal of pure inquiry and the autonomy it brings, these do not offer a clear and powerful case for public support of basic science.

This point needs to be strongly emphasized, lest the scientific community take false comfort from the new congressional leadership's greater support of basic science *relative* to the technology programs launched by the Bush and Clinton administrations. The danger here is to mistake relative for absolute favor. However reassuring it may be to find the new leadership embracing the economist's belief in the need for government to support basic research because the market will fail to do so, the fire will have gone out of this support if it is seen as directed only to *pure* research, just as a fire did not ignite the postwar support for basic science until the confrontation with the Soviets gave the policy community a clearer view of how vital this support might be for the country. In this regard, it should be noted that the neoclassical "market failure" argument for government support of basic science did not keep the House leadership from adopting a budget plan in the spring of 1995 that projected cumulative, constant-dollar reductions of 30 percent for basic science over a seven-year period.[7]

A policy based on the common recognition of the importance of use-inspired basic research could help break this impasse. Such a keystone to a renewed arch between science and government would be well matched to realities on both sides of this troubled dialogue. The realities of science include the fact that considerations of use have been part of the motivation for basic science since the late nineteenth century. The second industrial revolution was fueled by scientific discoveries inspired by industrial development, as the advances in public hygiene were fueled by the profoundly use-inspired discoveries of the scientist who has lent his name to this book. Indeed, in this century science has increasingly been furnished with problems for investigation by advances in technology and not only by the unfolding of its own inner agenda. The partial concealment of these realities by the paradigm view articulated by the Bush report has stayed the scientific community from revealing how much of its basic research is influenced by national need—and from helping to dispel the mistaken idea that the deepest scientific inquiry must be free of considerations of use.

In a complementary way, the realities of the policy community and public to which it is attuned include a considerable esteem for science as a force for human betterment. Absent the budgetary pressures that threaten to make a hostage of science policy, the policy community and broader public want to enlist the undoubted power of science to provide for unmet societal needs. The public opinion data are remarkably clear on the support for science in these terms, however distant and puzzling much of the public may find the scientific venture. The framework for the interplay of the policy and scientific communities that is provided by the attitudes and belief of the attentive public is of considerable importance in a democratic system with substantial tax monies committed to scientific research. The popular support for the federal investment in research and development was most authoritatively analyzed by Jon D. Miller and Kenneth Prewitt with support from the National Science Foundation more than a decade ago, and these studies have provided a baseline for Miller's contributions to the reports monitoring public opinion toward science that have subsequently appeared in the National Science Board's biennial *Science and Engineering Indicators* volumes.[8]

Following a conceptual approach pioneered by Gabriel Almond to chart public opinion in another area of limited public information and understanding, that is, foreign policy, Miller and Prewitt explored the information and attitudes about science and science policy held at the several levels of a "pyramid" whose layers, from apex to base, were made up of science policy decisionmakers, nongovernmental science policy leaders, the attentive public for science policy, the interested public, and the nonattentive public.[9] No one familiar with creationism or the assaults on science from such quarters as the animal rights movement and a wing of the environmental movement could miss the variety of the views of science held by the broad public. But the evidence built up by Miller and his colleagues over the past decade indicate that, however limited the information and understanding of science at the base of this pyramid, the general public's view of science is broadly favorable and that the information and support for science rise steadily at higher levels of the pyramid. Although the public is aware that science entails a degree of risk, Miller in 1981 found that the benefits of science outweighed the risks in the eyes of 90 percent of the attentive public, of 79 percent of the interested public, and of 66 percent of the nonattentive public.[10]

This research has also shown how deeply the public values science not for what it *is* but for what it's *for*. It is strikingly clear that the instrumental uses of science are the key to popular support. Especially the attentive public believes that "technological know-how" and "scientific creativity" are important sources of U.S. influence in the world and economic well-being at home, and they have high expectations for future achievements of science on such goals as a supply of cheap energy, a cure for cancer, and for the desalination of water. Miller summarized these findings by noting that

> the public attributes to science and technology a central role in the nation's influence in the world and in the standard of living of the American people. The level of expectations for future scientific achievements indicates that the past is viewed as prologue, especially among the attentive and interested publics. The substantial proportion of the public that reported that the benefits of scientific research had outweighed its risks points to a solid foundation of public confidence in science and technology.[11]

That this preponderant support has remained firm over the subsequent decade is illustrated by the results of a 1994 Harris poll that recorded 68 percent of a nationwide sample as agreeing that "science will solve many of the world's problems" and only 29 percent as disagreeing. This continued support is also attested by the indicators of popular attitude monitored by the National

Science Board, which include a 1993 survey showing that 79 percent of all adults in a nationwide sample agreed with the statement that "most scientists want to work on things that will make life better for the average person."[12]

In a period of uncertain party control and savage budgetary pressure, there are occasional calls from within the policy community for government to limit its support to the pure research envisaged by *Science, the Endless Frontier*. In the longer run, however, a broader understanding that a great deal of basic science is also inspired by considerations of use is likely to offer a far better hope of broad public backing. The first of three closely related reasons why a more realistic view of the relationship between advances in science and technology can strengthen the bridge between science and government can be stated as follows:

> The inspiration that basic research can draw from societal need strengthens its claim on public support in the policy community and from the public to which it responds.

Hence, from the perspective of the scientific community a wider, shared recognition of the importance of use-inspired basic research offers the possibility of strengthening the support for the fundamental research the scientific community wants to pursue in a period of fierce pressure on the federal budget. But this recognition can also increase the support for *pure* research, by an argument that invokes the unity of science and the unpredictable nature of scientific discovery. This argument leads to a second reason why a more realistic view of the relationship between advances in science and technology can help renew the compact betweeen science and government, since such a compact will not work if many in the scientific community feel that it will lead to a hemorrhage of resources from Bohr's to Pasteur's quadrants.

Strengthening the Case for Pure Research

In the traditional repertory of science policy, there is a well-rehearsed litany of reasons for supporting pure, curiosity-inspired research. A reason as old as the classical world is the belief that a civilized people will seek knowledge for its own sake. The belief in pure inquiry as a mark of civilization is influential in our own time as well, and the constituency for this belief is augmented by the popular curiosity about the unknown. The attentive public of a country that generates nine million subscriptions to *National Geographic* can be deeply interested in the revelations of the Hubble telescope about the origins of the universe—and find it reasonable for this curiosity to be satisfied at public expense, even though it finds other reasons for supporting science more compelling. Nothing in this analysis is meant to diminish the case for pure research in terms of the intrinsic, civilizing value of knowledge.

Yet Vannevar Bush found the appeal of knowledge for its own sake so unequal to the task of sustaining the flow of public support for basic science in peacetime that it went almost unmentioned in his report of forty pages, a document superbly tuned to the ear of his postwar audience. Bush instead centered his case on what has become the primary reason for supporting pure science—one dating from the Enlightenment—the belief that the advances in understanding achieved by pure research will later improve the human condition. This argument is central to the case made for basic science today, however much its force may be lessened by the doubts about the technology to which science has led or by doubts as to whether others may capture the return in commercial technology from our investment in pure research. Sometimes the proponents of pure science seek to counter the mixed feelings about the technological spin-offs of science by saying that we need even more knowledge to deal with these negative side effects. But this is clearly a call for something other than pure research, since the problems created by technology will in this case influence the renewed search for basic understanding—which becomes an exercise in use-inspired, rather than pure, basic research.

The case for investing in pure science has recently been buttressed by a variant of this argument—that the country needs reserve strength in all fields of science so that it will be ready to exploit unforeseen developments in technology unleashed by advances in some of these scientific fields. Just as it is hard to know what the technological spin-offs of science will be, so it is difficult to foresee what areas of science will undergird the development of future technologies. This argument was given considerable prominence by a report of the Committee on Science, Engineering, and

Public Policy of the National Academies of Science and Engineering and Institute of Medicine.[13] The need for such a hedge against an uncertain future lay behind the committee's proposal that "the United States should be *among* the world leaders, at least, in *all* major areas of science," a criterion that was spelled out in a subsequent lecture by Ralph E. Gomory, one of the report's principal authors.[14]

This argument reinforces the case for pure research by linking it to the country's need to remain competitive in an increasingly high-technology global economy. Only by being among the world's leaders in all areas of science will the United States be able to share in the unforeseen—indeed, unforeseeable—breakthroughs in technology to which the advances in some of these fields will surely lead. Without this investment, the United States could be shut out of these developments when they occur. Gomory illustrated his case with the experience of solid-state physics and molecular biology, fields in which he felt the scientific leadership of the United States allowed it to take advantage of unforeseen technological breakthroughs based on prior research in the purest realms of science.

This argument is a powerful restatement of the case for pure research in the circumstances of a competitive world. But it would be a pity for this argument to reinforce linear-model thinking by exaggerating the primacy of pure research. The scientists who in the 1930s created the field of solid-state physics as an extension of quantum mechanics had no practical purposes in view. But this field has exhibited a notably interactive relationship between science and technology in the decades since World War II.[15] The discovery of the transistor soon after the war established solid-state physics as a field of university study, thereby encouraging a good deal of pure as well as use-inspired basic research verifying and extending the quantum theory of solids. And the subsequent miniaturization of successive generations of semiconductors has in recent years inspired a number of condensed-matter physicists to seek the additional scientific understanding that will allow semiconductors to be built atomic layer by atomic layer. The interactive relationship of science and technology is even more marked in the field of molecular biology. The most practical of concerns led Oswald Avery to his discovery of DNA as the carrier of the genetic code, however much James Watson and Francis Crick may have viewed the problem of DNA's structure as a pure intellectual puzzle. The possibilities of biotechnology have never been far from the thoughts of those who have achieved the breakthroughs in molecular biology in subsequent decades, to a degree that almost overwhelmed the field when the methods of recombinant DNA were first discovered.

Indeed, the interactive nature of the relationship between science and technology in these and other fields offers a further argument for supporting pure research, one that supplies yet another reason for believing that the concept of use-inspired basic research can help renew the compact between science and government. This argument centers on the importance of strengthening pure research in those scientific fields in which the course of basic research bears on societal needs. The argument is lent its force by the unity of science and the profound uncertainty as to the future interplay of advances in scientific understanding and technological know-how. It is not the case that in a scientific field of demonstrated importance for social goals, the fundamental research that is influenced by societal need is one kind of science, while the fundamental research that is driven by curiosity alone is another. Both fall within a common scientific framework, however real the difference in goals that makes the work in Bohr's quadrant conceptually distinct from work in Pasteur's quadrant. Breakthroughs achieved by use-inspired basic research can lead to further *pure* research, just as breakthroughs in pure research can lead to further use-inspired research, all in accord with Pasteur's famous dictum that "there is not pure science and applied science but only science and the applications of science."[16] We may cite on the same point the words of Vannevar Bush and his colleagues on the panel appointed by Harvard's President James B. Conant to advise him on the use of the McKay Bequest:

A science, such as physics or chemistry or mathematics, is not the sum of two discrete parts, one pure and the other applied. It is an organic whole, with complex interrelationships throughout.[17]

There are limits to this argument for the spillover benefit for pure basic research from investments in use-inspired basic research. Work in fields of pure "intensive" science, with high-energy physics the paradigm case, will need to be justified in terms of the instrinsic value of knowledge and in two other terms as well. One is the value of innovation in the instrumentation developed for research in such fields, an argument that was legitimately, if ultimately unsuccessfully, advanced by advocates of the Superconducting Super Collider (SSC). The other is a human resource extension of the argument about the unity of science—the broad value for other fields of science and technology of having a set of researchers schooled in the most intensive fields of science. Harvey Brooks has observed that scientists trained in nuclear physics led the scientific success of World War II—on such problems as radar and the proximity fuse, as well as atomic weapons—and that scientists trained in nuclear physics, rather than solid-state physics, also led much of the early development of semiconductors. In his view, these "cross-overs" may have had less to do with the intellectual content of nuclear physics than with the familiarity with sophisticated electronic circuits and instrumentation and the systems-type thinking required by critical experiments in nuclear physics.[18]

Physicists trained in nuclear, or very fundamental atomic, physics—James Watson, Francis Crick, Wally Gilbert, and Seymour Berger—were equally conspicuous in the early development of molecular biology. But molecular biology is also a notable example of the case that the unity of sciences makes for investments in pure research once the case for investments in use-inspired basic research is clear. No one could doubt that the advances achieved by the pure science of Watson and Crick were as important as those of the use-inspired science of Avery, or of Herbert Boyer and Stanley Cohen, or of Georges Köhler and César Milstein in helping the field to realize the promise of biotechnology. Hence, the success of the research directed toward social goals broadened the support for the pure research that would strengthen the capacity of the field as a whole to meet societal needs.

There are many other cases of this mutual reinforcement in the annals of scientific research. In the history of organic chemistry, the breakthroughs by Herman Staudinger and Wallace Hume

Carothers ignited an explosion of further work that made polymer chemistry into a major field of chemical science. Some of this new work was influenced by the industrial value of synthesizing high-polymer materials, much as the work of these pioneers had been—with Staudinger supported by I. G. Farben AG and Carothers by Dupont.[19] But some of this subsequent research, largely university based, was a pure quest of the further understanding that helped give the field its intellectual structure. Both strands contributed to the capacity of polymer science to bring the world a remarkable diversity of new, high-polymer materials.

Such examples lead us to the second observation supporting the idea that a more realistic view of the relationship between advances in science and technology can help to strengthen the bridge between science and government:

> The societal value of use-inspired basic research within a scientific field strengthens the case for supporting the pure research on which the development of the field partly depends.

Hence, support for research inspired by societal goals need not divert support away from pure research or inspire a massive transfer of resources within a field from Bohr's to Pasteur's quadrant. On the contrary, as the emergence of goal-oriented basic research within a scientific field strengthens the case for public investment, it also strengthens the case for public investment in the pure research that will enhance the capacity of the field as a whole to meet the societal goals on which it bears. This point can be absorbed into the renewed compact between science and government if it is mutually accepted by the scientific and policy communities.

Capturing the Benefit in Technology

A shared recognition of use-inspired basic research can help strengthen the bridge between science and government in yet another way—by increasing the confidence that investments in this type of research will bring a return in technology to those who make them. This confidence sorely needs restoring, in view of how rapidly research knowledge can disseminate across national frontiers. Indeed, the pace of this diffusion has led a number of respected observers to wonder whether this premise of national investment in basic science continues to be viable. For example, Harold Shapiro, an economist prominent in the world of science and technology policy, has suggested that the rapid spread of scientific knowledge across national frontiers is diminishing the ability of single countries to pursue science-driven technology policies, just as the increasingly rapid transfers of capital had earlier diminished the ability of single countries to pursue independent policies of economic stabilization.[20]

So long as one clings to an older image of basic research as pure science it is appropriate to take a more pessimistic view of capturing the return from investments in basic science. Yet once the complementary idea is accepted of basic research that is use inspired, it is plausible to suppose that investments in basic knowledge guided by considerations of use are more likely to bring a substantial return to those who make them, whether use is defined in terms of economic gain or of other societal goals. The experience of a number of scientific fields supports the view that scientists involved in fundamental use-inspired research will play a role in the technological return from the resulting knowledge, enhancing the likelihood that the nation investing in the basic science will share in the technological return.

As discussed earlier, the stylized scientist who might be imagined from Bush's canons of basic research—an investigator remote from ideas of use whose curiosity leads to discoveries that only later provide the basis of new technology—offers a progressively less adequate image for modern science. By the late nineteenth century a number of scientists were exploring phenomena revealed by the advance of technology and were deeply involved in the technological return from the knowledge they gained. And, as we saw in prior chapters, this was by no means a passing phase in the development of science. Despite the power of the ideal of pure inquiry, equally notable examples arise in the twentieth century of basic scientists drawing their inspiration from applied needs and playing a role in the technological return from the knowledge they gain—as consultants, employees, entrepreneurs, or teachers and mentors of scientists who enter industry. As a result, those who have, for example, invested the funds through the National Institutes of Health and other agencies to support the massive buildup of molecular biology in the years since the Watson-Crick discoveries could have greater confidence that the country would capture a substantial share of the return in biotechnology as the advances in molecular biology established how DNA replicates itself, how the genetic code is read by living organisms, and how restriction enzymes can find genetic needles in chromosomal haystacks.

Hence a third general reason for believing that a more realistic view of the relationship between advances in science and technology can help rebuild the bridge between science and government may be stated:

> The uncertainty as to who will capture the benefit in technology from new scientific knowledge is lessened when basic research is directly influenced by potential use.

This observation can reduce the reluctance to invest in basic science that is felt by a policy community accustomed to the postwar view of basic research as pure research, driven by curiosity alone and leading to knowledge that is freely available to the technologically advanced countries.

Institutionalizing a New Compact

There is nothing self-exploiting about the opportunity to strengthen the bridge between science and government by coupling the inspiration basic science can draw from unmet societal need with the high value the policy community places on the problem-solving capacity of science. Although these interlocking realities enhance the promise of renewing the compact between science and government, this promise will be realized by strengthening the process for bringing together the two quite disparate kinds of judgments that shape agendas of use-inspired basic research— scientific judgments of research promise and political judgments of societal need.

After World War II this process question might have been explored in more general terms if the organizational plan proposed by Bush had prevailed and a National Research Foundation as broad as the wartime Office of Scientific Research and Development had inherited the government's investment in basic research. Such a foundation would almost certainly have addressed this process issue in broad terms as it built agendas of basic research in areas such as defense and health. But the defeat of this organizational plan gave much greater prominence to the distinctive paradigm view of basic science and technological innovation in *Science, the Endless Frontier,* which became a second line of defense against government control of the performance of research. Although such agencies as the Defense Department and National Institutes of Health developed considerable operational skill in building agendas of use-inspired basic research, the general conceptual question of how to gauge the potential of research for both understanding and use was largely swept under the carpet by Bush's canon that the attempt to mix these goals would be inherently self-defeating.[21]

Hence, it is hardly surprising that the leading scientific countries, equating basic with pure research, were slow to institutionalize a process for building agendas of Pasteur's quadrant research in the earliest postwar decades. This was as true of Britain and Germany as it was of the United States, although the Napoleonic tradition of purposive state action made the French a partial exception. This process issue has been considered in general terms only in more recent decades as a number of the OECD countries took a keener interest in "strategic research"—and with it, a keener interest in "research foresight." It should also occasion no surprise that the nation most alert to the value of developing a process for bringing together judgments of scientific promise and societal need was Japan, the country in which the radical separation of basic and applied research was least accepted.

The reasons why the Japanese have been the prime exception are cultural and historical. Less imbued by a rationalist scientific philosophy, they are perhaps also less inclined to such absolutist distinctions as the separation between pure inquiry and the practical arts that has had such a deep resonance in the Western tradition. Beyond this characteristic, the Japanese have from Meiji times believed that significant public and private investments should strengthen the country and safeguard its autonomy, and this impulse was heavily reinforced by the defeat in World War II. It was, as a result, natural for Japan to develop a process for linking scientific promise with societal goals as it made its investments in basic science.

Although each country must work out its own policy processes, it is odd that the United States should have been so inattentive to this aspect of the Japanese experience, in view of its awareness of Japan's industrial strides. Whatever its enthusiasm for Japan's quality circles or methods of reverse engineering and just-in-time assembly, America remained largely unaware of the Japanese efforts over several decades to develop a process of research foresight to guide their investments in research and development.[22] The foresight exercises periodically mounted by the Japanese government encompass both technology and science and are worldwide in scope; indeed, they are a natural outgrowth of the resolve to acquire the best of the world's technology that dates from Meiji times. In the postwar decades this scan enabled Japan to acquire a great deal of state-of-the-art technology, including much that was science based. As Japan became technologically and scientifically more mature, this search was linked to periodic reviews of Japan's investment in scientific research and technological development.

The Japanese have developed a considerable apparatus to conduct these periodic reviews. At the apex of the process is a Council for Science and Technology (CST), chaired by the prime minister and composed of very senior representatives of the major government ministries, of industry, and of the scientific community. The CST is assisted by a number of panels that include many additional senior figures from industry, government, and the scientific community. The eminence of these representatives gives the process great legitimacy in a consensual society and assures its conclusions wide influence. This structure does not, however, lead to a top-down, command-and-control process. Information about research "seeds" and social "needs" across a broad range of fields is elicited from scientists and engineers working at or near the bench and from many other informants by surveys, Delphi exercises, expert seminars, and specialized studies by research institutes. The process is an interactive one that allows informants to comment on the preliminary findings. The flows are managed by a separate division of the Science and Technology Agency, the most general scientific agency of the Japanese government. The most recent review, the eighteenth in a series extending back thirty years, culminated in the January 1992 publication of the CST's draft of a "Comprehensive and Basic Science and Technology Policy for the New Century," a document addressed by Kiichi Miyazawa, as chair of the council, to himself, as prime minister of Japan. The council's recommendations were accepted by the Cabinet in April 1992.

This process has had an important influence on both the micro- and macroallocation of research resources. At the retail level, it has encouraged firms to reexamine their research priorities against a backdrop of informed judgments about the promise of research in various subfields of science, requirements of the economy, and noneconomic societal needs. At a more global level, it has allowed private firms and public agencies to rethink their wholesale decisions about research support against this sort of backdrop.

Indeed, in the 1980s and 1990s these global priorities have increasingly focused on the place of basic research in the private and public investments in research and development. No strategic conclusion stood out more sharply from these reviews than the belief that Japan could no longer depend on acquiring and improving upon the world's technology. As clearly as their overseas critics, the Japanese understood how much of their economic progress since the war had involved a skillful acquisition of the world's technology and the scientific knowledge that underlay it. In a way reminiscent of Vannevar Bush's observation in the postwar years that America would now need to do its own basic research, the Japanese government accepted the conclusion of its Council on Science and Technology that the country needed to do more of its own fundamental science. This conclusion was reinforced by the awareness that Japan had come under increasing criticism for its deficit in the international trade in scientific ideas and would find it increasingly difficult to sustain a favorable business climate without making a substantial return to the world's stock of fundamental scientific knowledge.[23]

This impulse did not, however, translate mainly into increased investment in *pure* basic research, and the fact that it did not is relevant to our broader argument. With the well-developed sense of the importance of use-inspired basic research appropriate for a technologically oriented society, the government placed much of its own investment in basic science in research institutes with agendas of basic research that were strongly shaped by considerations of use and employed the indirect means at its disposal to encourage private firms to invest more heavily in basic research.[24] Associated with this authoritative statement by government was the rising interest in basic research in a number of Japan's largest business firms in the 1980s and early 1990s. Applying to themselves the conclusion reached by the Council on Science and Technology, an impressive group of Japan's flourishing, globally competitive companies established or strengthened basic research laboratories in this period. Hitachi, Toshiba, Canon, NEC, and others emphasized the commitment to research without expectation of immediate application in a way that echoed the earlier commitment of the GE of Langmuir, or the Dupont of Carothers, or the AT&T of Bell Labs' heyday. But the scientific phenomena explored in these laboratories were broadly related to technologies in which the firms were strong; indeed, phenomena that were often *revealed* by these technologies.[25] Therefore Japan's increased investment in basic science has been substantially devoted to Pasteur's quadrant rather than to Bohr's quadrant research, and the government's periodic foresight reviews are partly to be understood as efforts to bring together assessments of research "seeds" and of societal "needs."[26]

Although no nation's experience can serve as a template for others, the manner in which Japan has coupled judgments on "needs" and "seeds" as it builds agendas of use-inspired basic research deserves to be heeded elsewhere. The Germans have funded an adaptation of the Japanese "foresight" methods.[27] Partly inspired by the Japanese, Britain's Office of Science and Technology has followed up on the Government's White Paper on science and technology with fifteen "foresight" reports on the research and development needs of major sectors of industry.[28]

This review of Japanese practice has been useful mainly to underscore the need to rethink the way agendas of use-inspired basic research are shaped in the United States, a subject explored in chapter 5.

5 | BASIC SCIENCE AND AMERICAN DEMOCRACY

IN THE SETTING of American democracy, a broad awareness of how deeply modern science is inspired by societal need is more likely to renew the compact between science and government than is a generalized promise of a technological return from *pure* science. So runs the argument of chapter 4. Even in the years following World War II the promise of pure science as a remote but powerful dynamo of future technology was not enough to open the federal purse strings until the Soviet challenge gave the policy community a better understanding of what basic research was *for*. Only with the launching of *Sputnik* a dozen years after publication of *Science, the Endless Frontier* was the postwar bargain between science and government properly struck.

Nor should the scientific community take a new congressional leadership's acceptance of the need to support basic science as a public good to be a conversion to Vannevar Bush's case for pure research. This economist's argument tells us why the market will not supply a public good; it tells us nothing about what the good is worth. If the R&D expenditures of the federal government were released from the vise in which all discretionary spending is presently held, more than the generalized promise of a return in technology from investments in pure science would be needed to put the support of basic research on a strongly upward course.

The risk that an explicit recognition of the importance of use-inspired basic research will encourage the government to erode the autonomy of basic science is much less severe than it seemed to Bush and his colleagues when they set out his case for pure research at the end of the war. Over the postwar decades the country has evolved a diversified system of federal support for basic science, one that includes a great deal of basic research that is partly inspired by societal need, broadly conceived. In a later section of this chapter the conceptual lens of chapter 3 is used to describe how the federal government has evolved substantial agendas of Pasteur's quadrant research as the mission agencies extended their R&D investments beyond Edison's quadrant and the scientific R&D agencies, especially the National Science Foundation (NSF), extended their investments beyond Bohr's quadrant.

But the influence of the postwar paradigm has impeded these developments at many stages along the way. The conceptual lens of chapter 3 is also used to trace the unfortunate influence of a paradigm view that continued to partition the world of research into pure science and purely applied science. An intellectual system that concealed the presence of Pasteur's quadrant has made it far more difficult to address in general terms the central question raised by this final chapter—how judgments of scientific promise and of social value can be brought together in federal investments in use-inspired basic research.

A more realistic paradigm view of basic science and technological change can help to clarify this problem of science policy, one that is of broad importance for basic science in the setting of America's system of political democracy. In the following section the contrasting roles of the scientific and policy communities in applying scientific and use criteria at the "retail" level of funding individual projects through systems of peer review are explored. The distorting effect of the postwar paradigm on the evolution of "wholesale" support of categories of use-inspired basic research is also traced. Finally, several unresolved issues of science policy, which a more realistic paradigm view of basic research can help to clarify, are reviewed. These issues are of substantial importance for the health of basic science in the setting of American democracy.

We can make a start toward understanding how a system of allocation can take account both of scientific promise and of social value by focusing on the fundamental distinction between the funding of basic research at the project level and the funding of basic research at a more aggregative level. The needs of an allocation system are quite different along this "vertical" dimension extending upward from the retail choices about funding individual projects to wholesale decisions about funding programs or categories of basic research.[1] We will see that the task of combining informed judgments about scientific promise and societal need is transformed between these *micro*allocative and *macro*allocative levels of choice.

Recognizing Scientific Promise and Social Value at the Project Level

From the definition of use-inspired basic science it is clear that the decision to proceed with a particular project involves a judgment both of scientific promise and of social value. A key to understanding how these criteria can best be applied at the project level is to note that judgments of social value will typically involve one or more goals that do not require deep technical background to comprehend, while the judgments of scientific promise will typically require professional expertise that is held only by scientists with considerable background in the area of research. Hence, there is typically a marked asymmetry in the difficulty of judging social value and research promise at the project level—and, by extension, a marked asymmetry in the information that scientists with mastery of a field and those without this mastery can bring to the choice among alternative projects.

These asymmetries may be clearer if we consider a prototypical example of research in which the quest of fundamental knowledge is partly inspired by considerations of use. Suppose that a protein chemist is renewing the quest for the scientific understanding that will allow the synthesis of blood by developing a substitute for hemoglobin capable of transporting oxygen. This quest was given an enormous forward thrust earlier in the century when Max F.

Perutz earned a Nobel Prize for transforming the general under-standing of proteins by using the technique of X-ray diffraction to "solve" the structure of the hemoglobin molecule, one with more than 5,000 atoms. There is little doubt that Perutz turned his crystallographic studies to this molecule partly because of its im-portance in human blood, and several decades later the immense potential market for improved blood continues to inspire research into the relationship between the structure and function of the hemoglobin molecule. Indeed, our prototypical scientist might be pursuing research within one of the companies that have been formed to enter this race. But he or she might equally be a univer-sity-based chemist, alert to the scientific and societal importance of this research, who is seeking funding from the National Insti-tutes of Health (NIH).

What is to be learned by considering the situation of this pro-totypical investigator? A first elementary observation is the con-trast between the conceptual difficulty of the scientific and use criteria that are involved. The project's applied goal will in general be much easier to grasp than the theoretical and technical elements of the field of protein chemistry that shape the scientific objectives and design of the research. The more fundamental the research, the likelier this is to be true.

A further observation follows directly from the first: that there is an inherent contrast between the relative ease with which sci-entists with technical knowledge of a field can grasp its applied goals and the far greater difficulty of those outside this circle in appraising a project's scientific promise. This contrast is evident even among scientists who are closely familiar with the field in question and scientists from more distant fields. But the contrast is very marked indeed between scientists who have mastered the field and those without scientific training, even if they hold posi-tions of political authority within the allocative system. The sci-entist and lay observer are on vastly different footing in judging the scientific promise of next research steps in understanding the properties of hemoglobin. But the protein chemist and his or her funders are on much more equal footing in knowing the value of synthesizing blood free of HIV virus in a world suffering from AIDS.

This conclusion ought not to be pressed too far, since the analysis of social value may in some cases involve considerable technical rigor. If, for example, social value resides in commercial profitability, nonscientists may be far more adept in the analysis of prospective revenues and costs that is required to say what this value may be. Similarly, nonscientists may have an advantage in the analysis of complex moral and ethical issues that is sometimes required to judge the value of a potential application of scientific knowledge. But these are exceptions; the appraisal of scientific promise will typically be far more demanding than the understanding of social value at the project level.

Three implications of broad significance flow from these observations about the relative ease of gauging the scientific and social value of research. The first is that the scientists embarked on Pasteur's quadrant research can play an important role in clarifying the potential social value of their work. In many cases, societal needs have partly inspired their work, and the conduct of their research can involve an interactive growth in their understanding of the scientific structure of the problem and the societal needs on which it bears. The annals of research are richly supplied with examples of this process. Some were indeed furnished by the scientist whose name appears in the title of this book. As his research provided an adequate theory of the problem that had stymied the efforts of the public hygiene movement in England and France, Louis Pasteur laid out a whole new agenda of public health needs and opportunities, such as purifying milk that was infecting children with tuberculosis. A contemporary example is supplied by the studies of the effects of chlorofluorocarbon (CFC) gases in the upper atmosphere, a case in which the initiative in defining societal need again came from the scientific community. It required the extraordinary scientific detective work for which Mario Molina and F. Sherwood Rowland received the Nobel Prize in chemistry in 1995 to make the world aware of the need to prevent the depletion of the ozone layer if it were not to risk substantial plant and animal damage and rising cancer rates. The initial obliviousness of the policy community and public to this social need gave way to cautious awareness and then to broad acceptance as the predictions of Molina and Rowland were confirmed by observations of the ozone layer made by the National Aeronautics and Space Administration (NASA) and by a British research group working in Antarctica.[2]

The importance of the creative involvement of scientists in defining societal need deserves some emphasis, in view of how effectively the postwar paradigm has obscured this phenomenon as well. As noted in chapter 3, it was specifically downplayed by Alan T. Waterman and others who sought to rescue Bush's canon, that basic research is performed without thought of practical ends, in a period when it was becoming increasingly difficult to overlook the influence of applied goals on the course of basic science. As Waterman laid out the concept of "mission-oriented" basic research, only the funders of research would be concerned with considerations of use, while the bench scientist remained free to pursue his or her fundamental inquiries without thought of practical ends. The first of our implications may be stated as follows:

A system for appraising scientific promise and social value at the project level should enlist the insight of the working scientist into the nature of the social goals on which his or her research bears.

A second implication that flows from the relative ease of gauging scientific and social value at the project level is that it is unwise to use a system of allocation that separates these two judgments. There is a surface appeal to the idea that they ought to be assigned to those most competent to make them, with a project's research promise being judged by scientists close to the field of research, while its social value is judged by those with organizational or political authority to define social goals. This appeal has been strong enough to inspire recurring efforts to institutionalize such a division of labor. Whatever its surface logic, however, this arrangement loses the creative insight of the bench scientist in helping to define societal need. It also runs the risk of creating a bifurcated funding system with substantial conflict between those who are asked to judge scientific promise and those who are asked to judge social value.

These disabilities are illustrated by a famous episode of science policy in Britain. In the early 1970s a plan for giving separate and equal weight to judgments of research "seeds" and societal "needs" was proposed by Lord Rothschild when he served as chief of the Central Policy Review Staff under prime minister Edward Heath. A "customer-contractor" principle was the keystone of the Rothschild Report and was accepted by the Government White Paper on Science in 1972.[3] Under this principle, funds were to be awarded projects with applied goals by joint action by one of the science research councils, which would judge their scientific promise, and by one of the government departments declared in those pre-Thatcher days to be the "customers" who would judge the social value of the research.

The relevance of the Rothschild Report and Government White Paper for our argument is obscured by Rothschild's total acceptance of the postwar dogma of the separation of basic and applied research. Indeed, the science council most hostile to the Rothschild plan, the Medical Research Council (MRC), was equally hostage to the postwar paradigm and argued its case in terms of the value of fundamental science, since it too lacked a conceptual vocabulary transcending the basic-applied dichotomy. But its argument was really about the funding of use-inspired fundamental science. The MRC ultimately prevailed by convincing its "client," a remarkably enlightened Department of Health and Social Security, that the funds in question would have far greater impact on the nation's health if allocated by the MRC to the most scientifically promising, yet typically use-inspired, studies proposed to the council. The MRC's Cambridge Laboratory of Molecular Biology was subsequently the site of such studies as Aaron Klug's research on the tangled aggregations of molecules that lead to the death of nerve cells in Alzheimer's disease, as well as Georges Köhler and César Milstein's Nobel Prize–winning achievement in harnessing cancer cells to produce monoclonal antibodies in great quantity, a breakthrough that, together with the development of the methods of recombinant DNA, revolutionized the fields of molecular biology and biotechnology.

The third implication to flow from the relative ease of understanding the scientific and social goals of research is the ability of the bench scientist to comprehend and accommodate to the incentives having to do with societal need that may be built into a system of funding research at the retail level. In particular, experience has shown the feasibility of introducing such incentives into systems of peer review, which are often seen purely as a means of judging the scientific promise of alternative projects. This point is supported by the comparison of practice in the National Science Foundation, where peer review is almost wholly dedicated to gauging research promise, and in the National Institutes of Health, where peer review has been administered with an eye to the country's medical needs since the interwar years.

Peer review has a considerable history in scientific *publication*; the logic of such a process for allocating scarce pages in scientific journals was, as Robert K. Merton and Harriett Zuckerman note, evident from the dawn of the Royal Society in Restoration England.[4] Although peer review of journal articles has long been accepted, a *grants* peer review process evolved at the fledgling NIH, especially the National Cancer Institute, toward the end of the interwar period, and this device became a cornerstone of the process by which NIH selected projects for support in the postwar years. A somewhat different process of the microallocation of research support through peer review grew up in the National Science Foundation, from a seedling brought to the Foundation from the Office of Naval Research by Waterman, NSF's first director.

Although these NIH and NSF models of grants peer review are very similar, the differences between them reflect the greater early commitment of NSF to pure research. Each of the proposals to a given NSF program is typically sent out for review by a number of scientific peers judged by the program officer to be best able to assess the scientific excellence of the proposal and the qualifications of the investigator. These reviews strongly influence the program officers and the disciplinary panels that advise a number of the programs, although the Foundation's officers are ultimately responsible for making awards. Such reviews are often supplemented by site visits in the case of large projects and special competitions.

By contrast, NIH initially relies not on mail assessments by peer scientists who are intimately familiar with the technical content of a subfield but on the assessments reached by reviewers who make up more than 100 study sections of scientists outside of the government who meet several times a year on NIH's Bethesda campus to assess the proposals assigned to their sections. For each proposal one member of the study section serves as the primary reviewer and one as the secondary reviewer, chosen for their ability to gauge the scientific quality of the proposal and applicant. Each writes a detailed critique. A third member serves as a reader who is especially prepared to discuss the proposal when the study section meets.

The NIH procedure is therefore less deep in terms of the judgments of peers most knowledgeable in the area of a particular proposal but provides a more comprehensive view of how well the science funded by a set of awards will serve the biomedical goals for which the study section is responsible, and the influence of programmatic considerations becomes still more explicit when the recommendations of the study sections are reviewed by the legislatively mandated National Advisory Councils, one for each institute of NIH. These councils, consisting of scientists and "community leaders in areas relevant to the health areas and scientific responsibilities of the particular Institute," approve the funding of proposals on the basis of program goals and priorities as well as scientific merit.[5] This system of project review has been an essential element at the retail level of NIH's extraordinary success in developing a "comprehensive" strategy of wholesale allocation that encompasses work in Bohr's, Pasteur's, and Edison's quadrants but is fairly clearly centered on Pasteur's quadrant. This strategy is discussed in the next section, when the evolution of the federal government's macroallocative policies of research support in the postwar decades is described.

Peer review is by no means the universal method of allocating federal support for use-inspired basic research at the project level. Although it is used by a variety of mission agencies and the Office of Management and Budget and Office of Science and Technology Policy are strongly encouraging its spread, it is less well suited to the big science projects that are undertaken by many of the federal government's own research establishments. The national laboratories typically use quite different means of allocating research funding at the retail level.

Moreover, from the moment of its inception, grants peer review had its critics, who over several decades have produced a considerable literature in the letters columns of *Science* and other journals and in occasional more substantial pieces.[6] The critics have said that peer review is biased toward existing rather than novel approaches, toward higher-profile rather than lesser-known researchers, toward projects within the established disciplines rather than across disciplines, toward research in high-prestige rather than less prestigious institutions. Its survival they attribute to the inevitable satisfaction of scientists who have done well under the system, suggesting that peer review is yet another realm of human experience subject to Abraham Lincoln's taunt that those who like this sort of thing will find this the sort of thing they like.

The response of the agencies administering peer review, with prods from Congress, has been a series of studies assessing the performance of the system and recommending perfecting changes.[7] One change was a shift of nomenclature from "peer review" to the more easily explained "merit review." Although these reports have failed to silence the most articulate critics, they bear out the broad conclusion that peer review is a reasonable way to allocate funds to individual basic research projects according to judgments of scientific promise, one that stands up well against the alternative methods of allocation that have been proposed. Indeed, the scientific community's current complaints about peer review are largely provoked by the low rate of yield of the grants process and by the realization that peer review has a far more difficult task when it is asked to decide among a plethora of excellent proposals. This extrinsic disability of grants peer review will be still more troubling in a period of severe budget stringency, and there is unease that investigators may write proposals defensively rather than imaginatively, lest a review panel seize on an element of weakness to simplify an impossible task. NIH in particular is considering ways of adapting its review procedures to the realities of a period of drastically reduced success rates.

All the same, microallocation by peer review has almost certainly played a major role in the extraordinary achievements of federally funded basic science over half a century, including the achievements of such use-inspired areas of fundamental research as biomedicine and health. If the return on projects judged most promising in research terms is only marginally greater in a given period, such an increment compounded over time can lead to a markedly greater yield from the investment in basic science. This conclusion deserves to be forcefully stated as the conception of basic science is broadened to include fundamental research that is also inspired by applied ends. Imbued with an older, linear-model thinking, parts of the policy community may want to see basic yet use-inspired work simply as applied research and to suppose that a different kind of microallocation is appropriate; for the U.S. Congress, a superlatively geographic institution, a natural alternative is to distribute research funding by geographic formula.[8] Echoes of this view could be heard in the remarks of a congressional leader, a long-standing friend of basic science, who told the January 1994 Forum on Science in the National Interest convened by the Office of Science and Technology Policy that if science was now to help create jobs in the U.S. economy, his economically hard-pressed state should have a share of the funds for *performing* the research. This geographic impulse is also clear in congressional "earmarking" of funds for research infrastructure.

Therefore, the conclusion that is suggested by the contrast between microallocation by peer review and by alternative methods may be stated as follows:

> Public funds invested in use-inspired basic research will bring a greater return if they are allocated among alternative projects through peer review by panels capable of judging scientific promise and the societal benefit from the resulting scientific knowledge.

This conclusion was endorsed by the memorandum on Fiscal Year 1996 Research and Development Priorities jointly issued by the director of the Office of Management and Budget and the director of the Office of Science and Technology Policy, a circular that has quickened the interest in peer review in a number of the federal mission agencies that had utilized other means of funding basic research.[9]

Linking Scientific Promise to Social Value at the Wholesale Level

The problem of building and allocating funds to agendas of use-inspired basic research is fundamentally transformed as one moves from the microallocation of resources to the macroallocation of resources to categories of research. Both the scientific and policy communities have at the wholesale level of choice different roles. No longer can the leading scientists in various fields, especially those where the frontiers of knowledge are pushed back mainly by "small science" projects lending themselves to peer review, make judgments about scientific promise applying a deep knowledge of a particular field. The broader the categories of research to which funds are allocated, the greater this transformation will be.

The policy community too has a markedly different role as one moves from the retail to the wholesale level of research funding. In America's governmental system, judgments of social value will ultimately be made by the cut and thrust of pluralist democracy, in which those in Congress and the White House who are clothed in democratic authority by election to office have the most influential role. Those wielding this authority will directly determine the fate of individual big-science projects, as they settled the fate of the star-crossed Superconducting Supercollider project in high-energy physics. By contrast, the role of the policy community will typically be limited to setting overall budgetary levels in the case of small-science projects that lend themselves to peer review—and to giving broad legislation on budgetary sanction to the societal needs that may partly guide the peer review process, such as the health goals that partly guide decisions of many of NIH's study sections and National Advisory Councils.

In this section the problems of institutionalizing the role of the scientific and policy communities in blending scientific promise and social value at the wholesale level of research funding are revisited. The conceptual framework of chapter 3 is used to show how the federal government built a series of ad hoc policies for supporting use-inspired basic research in the postwar decades and to show how deeply the postwar paradigm compounded these developments.

If one examines the evolution of federal support of basic science through the lens of chapter 3, three developments by which the federal government has implicitly sought to build and fund agendas of use-inspired basic research in the postwar decades can be distinguished. The first has involved an increasing recognition of the use-inspired elements within the scientific and engineering fields by the National Science Foundation, the hallmark institution of pure research, a development that might be said to have extended NSF's portfolio of research from Bohr's quadrant into Pasteur's quadrant. The second has involved the recognition of the importance of a more fundamental understanding of scientific phenomena bearing on the goals of the mission agencies of the government, a development that might be said to have extended the R&D programs of these agencies from Edison's quadrant to Pasteur's quadrant. The third has involved NIH's recognition from the earliest postwar years that it could best fulfill its mission by building extensive agendas of use-inspired basic science as well as by funding a considerable volume of pure basic research and of purely applied research. This third pattern might therefore be said to involve a more comprehensive portfolio of investments centered on Pasteur's quadrant.

Extending Support Beyond Bohr's Quadrant in the Practice of NSF

By creating the National Science Foundation five years after the publication of *Science, the Endless Frontier*, the federal government provided an explicit channel for funds dedicated to basic research. In view of the investments in basic science then being made by the Defense Department, the Atomic Energy Commission, and the National Institutes of Health, this new channel was not the sole or even the most important route for funds to basic research, although it became impressively broad and deep in the aftermath of *Sputnik*, and was from the beginning distinguished by the quality of the science it supported by peer review. It was expected that this channel would be almost entirely dedicated to the pure research anticipated by the Bush report; in the 1950s and 1960s, NSF understood its primary role to be the support of curiosity-inspired disciplinary research in the universities, and this Bohr's-quadrant mission was vigorously defended by NSF's academic constituency.[10]

In terms of the criterion of scientific promise, NSF has faithfully adhered to this mission in the years since; if the conceptual plane of chapter 3 is again invoked, the Foundation has made no concession on its commitment to invest in research high on the vertical dimension of fundamental understanding. Over time, however,

NSF has taken a more open view of where to place its investments along the horizontal dimension representing the influence of applied goals on projects of research. An undoubted factor in this shift has been the broadening of the Foundation's scope to include the engineering fields; the agency was bound to be moved toward a more open stance on research that is inspired by societal need as it responded to the successful, several-decade campaign of the engineers to join the scientists under NSF's tent. (This was itself a fascinating chapter in the assertion of professional interest culminating in the 1986 amendment to NSF's Organic Act that accorded engineering equal status with science.)[11] But the engineering disciplines were by no means the only fields within NSF's scope where research could be influenced by applied goals, as the review of the interventionist and observational sciences in chapter 1 made clear. An increasing awareness of the interplay of understanding and use in biology and the other life sciences, in physics and chemistry, and in geology, atmospheric, and oceanic science and other earth sciences helped move the Foundation to a more accepting attitude toward considerations of use.[12]

Other organizational initiatives reinforced this shift. None was more impressive than NSF's success in the late 1970s in harnessing the disciplinary strengths of physics and chemistry, chemical and electrical engineering, and metallurgy and materials science to establish new Materials Research Laboratories. These laboratories were partly inherited from the Advanced Research Projects Agency (ARPA) of the Department of Defense, when the Mansfield amendment (which required that the Defense Department's basic research be clearly related to its mission) led OMB to encourage the transfer to NSF of ARPA's fundamental research. NSF's Materials Research Laboratories anticipated the Engineering Research Centers and the Science and Technology Centers created by the Foundation from the mid-1980s, launched during the directorship of Erich Bloch, a distinguished engineer with a background in industry.

This shift had carried far enough by the 1990s for a special National Science Board of the NSF Commission on the Future, appointed during the directorship of Walter Massey, to incorporate in its general recommendations the observations that

the history of science and its uses suggests that the NSF

should have two goals in the allocation of its resources. One is to support first-rate research at many points on the frontiers of knowledge, identified and defined by the best researchers. The second goal is a balanced allocation of resources in strategic research areas in response to scientific opportunities to meet national goals.

It is in the national interest to pursue both goals with vigor and in a balanced way.[13]

Senator Barbara A. Mikulski, who chaired the Senate appropriations subcommittee responsible for NSF's budget, was quick to understand "in a balanced way" as meaning "to give equal weight to" when Massey subsequently testified on NSF's fiscal year 1994 budget. Massey responded by saying that the Foundation ought "to harness the strengths of the Nation's research enterprise on broad strategic goals" and agreed in budget terms that the division "will be about 50/50, in fact, if we track the funds."[14] The subcommittee promptly increased the pressure on the Foundation by including in its report that "not less than 60 percent of the agency's annual program research activities should be strategic in nature."[15] At a number of points, however, the mindset of those imbued with the postwar paradigm impeded the broadening of the NSF's portfolio to include use-inspired basic research.

We have seen how vigorously NSF's academic constituency, deeply persuaded that research beyond the realm of pure science is purely applied, defended the primacy of the Foundation's mission in curiosity-driven, disciplinary research in the Foundation's early years. But the policy community, having also absorbed the idea that basic and applied research are necessarily separate, only reinforced the scientists' fears that whatever was added to this traditional research mission must be purely applied. Such an ironic scenario was played out in the passage of the Daddario amendments that broadened the Foundation's charter in the late 1960s, as the policy community sought to harness the power of science for social goals. Accepting the dogma of basic and applied as separate realms, the congressional sponsors of these amendments supposed that the appropriate course was to incorporate applied research into NSF's charter rather than to encourage the Foundation's support of fundamental research that is inspired by societal need. With the amendments to NSF's charter conceived in these terms, the Foundation's academic constituency understandably saw these new activities as lying beyond the realm of basic science and in competition with basic research for scarce funds.

A clearer view of the distorting role of the postwar paradigm can be seen in the fate of RANN (Research Applied to National Needs), NSF's programmatic response to Congress's adoption of the Daddario amendments. This was a time at the juncture between the Johnson and Nixon administrations when Congress and the executive branch shared a concern as to whether technology transfer was bringing the country an adequate return from its investment in science—through the National Institutes of Health, the Department of Defense, and the National Science Foundation. Both the outgoing and incoming presidents showed their impatience with this return, and President Richard M. Nixon soon announced his War on Cancer.

How NSF should respond to its broader charter was a matter of considerable tension within the Foundation.[16] An activist stance was advocated by the Foundation's director, William D. McElroy, and a number of his lieutenants, as well as by spokesmen for the engineering community who were pressing for a larger role for engineering within the Foundation. McElroy's incentives for such a stance were greatly strengthened in December 1970 when the Office of Management and Budget offered to increase the Foundation's fiscal year 1972 budget by $100 million, rather than by the $13 million it had modestly requested, if it would put together a new program to enlist science and technology in the nation's needs. For a leadership intent on making NSF a billion-dollar agency, the lure of such a proposal was understandable.

But a number of scientists within NSF's board and research directorate viewed the broadening of the Foundation's mission to include applied science with a hostility reminiscent of Bush's variant of Gresham's Law that applied would always drive out pure if the two forms of research were mixed. The National Science Board noted its fear that over time Congress's enthusiasm for applied science would lead to a "relative diminution in the basic science portion of the [NSF's] budget."[17] Although they were unable to block a major initiative to respond to the challenge by Congress and OMB, they created so hostile a climate for such an initiative that it was exceedingly difficult for the Foundation to exploit the opportunity it faced.

Under fairer skies the Foundation might have taken the broadening of its congressional charter and the prospect of further funding as an occasion for radically rethinking the separation of basic and applied research and investing the new funds in areas of basic science where research was directly influenced by considerations of use. There are glimmers of such a vision in the justification for NSF's earlier initiative, IRRPOS (*Interdisciplinary Research Relevant to Problems of Our Society*), by which it sought in 1968 and 1969 to respond to the Daddario amendments. The logic for rethinking the separation between basic and applied research in a larger program was strengthened by OMB's stipulating that half of NSF's new funds be used to support basic research cut loose from the mission agencies.[18] McElroy claimed that as much as 40 percent of the funds for his fresh initiative would be devoted to basic research, and he embraced the acronym RANN (*Research Applied to National Needs*) as implying "that either basic or applied research would . . . be directed to the solution of identified problems."[19]

It is hardly surprising that the hostility of NSF's science constituency, augmented by doubts in the National Science Board about whether NSF had a comparative advantage over the mission agencies in undertaking this type of research, kept the Foundation from seizing this opportunity. Notably absent was the rationale that the concept of use-inspired basic research might have provided for distinguishing the NSF's from the mission agencies' role in research influenced by considerations of use. The budget justification and programming strategies for RANN prepared at breakneck speed by an eight-person task force chaired by Joel Snow, IRRPOS's director, were mainly focused on applied research on the environment, urban problems, and the country's energy needs. Almost from the beginning RANN was also hobbled by the prospect, in accord with good linear-model thinking, that NSF would be a catalyst and at the "proof-of-concept stage" would transfer to appropriate mission agencies or to private industry the findings and responsibility for the applied research it started. In 1975 this promissory note came due when NSF transferred to the newly organized Energy Research and Development Administration

(ERDA) virtually the whole of the solar and other energy projects that had been given a strong thrust forward by the oil shock of 1973. This end of the "walk over the energy mountain" went unlamented by NSF's basic science constituency, and in 1977 the National Science Board and a new NSF director terminated RANN as abruptly as it had been launched six and one-half years earlier. The 1994 admonition of Senator Mikulski's subcommittee that not less than 60 percent of the NSF's annual program research allocation should be strategic in nature still sent a shudder through parts of the scientific community that were accustomed to the basic-applied distinction and were therefore quick to associate "strategic research" with narrowly targeted applied work.

Extending Support from Edison's Quadrant in the Practice of the Mission Agencies

A small but telling mark of the triumph of scientific outlook in modern America is the routine practice of incorporating authority to undertake research and development that bear on the agencies' mission in the legislation authorizing the agencies of the federal government. Most of the research supported by the vast resulting array of R&D accounts is purely applied and closely aligned with the operating needs of the mission agencies. And in the early 1970s the Mansfield amendment settled a chill on basic science in the mission agencies; even if its requirement that basic research be clearly related to an agency's mission applied only to the Defense Department and only to a single fiscal year, a number of other agencies accepted the either-or logic of the basic-applied distinction and felt that they too should limit their R&D activities to purely applied research and development lest they come under similar fire.

There were, however, notable exceptions to this reluctance among the mission agencies to extend their R&D activities beyond Edison's quadrant, since it was clear that some agencies could achieve their goals only by promoting a more fundamental understanding of the scientific phenomena that bore on their missions. An early case was furnished by the U.S. Department of Agriculture, although USDA's role in basic research has receded in recent years. America became the pacemaker of world agriculture by drawing on the knowledge gained from the federal investment in agricultural science that goes back to the founding of the land grant universities and agricultural experiment stations in the nineteenth century. Although a great deal of the R&D funded by the Agriculture Department too is purely applied, the department's budget for fiscal year 1996 contained $605 million for basic research, including a number of fundamental studies in fields such as molecular biology.[20] There is, in effect, substantial consensus among the relevant parts of the policy community, its agricultural constituency, and a community of agricultural scientists that the department's mission requires the extension of its R&D programs upward from Edison's to Pasteur's quadrant.

Four other mission agencies of the government surpass the volume of basic research in the Agriculture Department. Each has a distinctive background to its willingness to extend its R&D activities beyond the purely applied. In the early years of the cold war the service departments moved swiftly to promote the flow of basic research that bore on the country's security without waiting for the emergence of a unified Department of Defense. Indeed, we have seen how far the Office of Naval Research went toward becoming a National Science Foundation-in-waiting by funding a large volume of research in Bohr's as well as Pasteur's quadrant to help fill the gap created by the failure of Bush's organizational plan. In the postwar years the Defense Department was also willing to buy into fundamental Pasteur's quadrant research, as it did, for example, in funding John Slater and Arthur von Hippel's pioneering research in condensed-matter physics at MIT, research that helped lay the foundation for the postwar boom in materials science.[21] Although there were periods of ebb as well as flow in the military's enthusiasm for basic research, the budget for fiscal year 1996 included $1.214 billion for basic research sponsored by the Defense Department.

The Energy Department's portfolio of basic research—totaling $1.785 billion in fiscal year 1996—has been assembled from two main sources. One is the interest in alternative energy sources inspired by the oil shocks of the 1970s, an interest still highly relevant to the country's needs. The other is the research of the Atomic Energy Commission, for which DOE inherited responsibility (via the Energy Research and Development Administration) when the reluctance to have the AEC serving as both developer and regulator of nuclear energy led to its dismemberment in 1974 and the assignment of the first of these roles to the Energy Department, the second to the Nuclear Regulatory Commission. Because these transfers have weakened the postwar belief that the research underlying atomic weapons should be in civilian hands, it is now sometimes thought to be an anomaly for the Energy Department, rather than Defense, to operate the nation's weapons laboratories. Another anomalous legacy from the AEC is DOE's sponsorship of the remarkably pure research in high-energy physics, a link explained by institutional history and not at all by the accidental presence of "energy" in the name of the field.

Special characteristics mark the R&D programs of the two other mission agencies with budget authority for basic research exceeding the Department of Agriculture's. One of these is the National Aeronautics and Space Administration, where a substantial fraction of the funds for basic research ($1.841 billion in fiscal year 1996) is devoted to operational projects. The other is the National Institutes of Health, administratively located within the Public Health Service and the Department of Health and Human Services. NIH is distinctive not only for the enormous scale of its support of basic research, at $6.311 billion in fiscal year 1996 by far the largest among the federal agencies, but for the degree to which this support is centered on use-inspired basic science. Indeed, NIH deserves to be seen not as a mission agency but as a scientific R&D agency that has pioneered a comprehensive strategy that spans Bohr's, Pasteur's, and Edison's quadrants.

The postwar paradigm's concealment of use-inspired basic research has introduced a different set of distortions into the development of basic science in the mission agencies. With the scientific and policy communities believing that basic research is performed without thought of practical ends, mission agencies that nonetheless wanted to involve first-class scientists in their problems did so by cobbling together offers of support for the investigator's own basic research, as pure as the investigator wished it to be, in exchange for applied research and consulting of importance to the agencies. In the early postwar years, there were examples of this stratagem at both the retail and wholesale levels of research support. At the retail level, the Defense Department achieved such an amalgam in a variety of its contracts with individual scientists and engineers by adding an override to support the investigator's own research. These arrangements almost certainly helped a generation of physical scientists to perceive themselves as pursuing free research even if their funding came from the Defense Department, a self-perception that was increasingly important as the controversy over the department's funding erupted on university campuses during the Vietnam War. Long after the year when the Mansfield amendment legally constrained the Defense Department's budget, this magnificent product of linear-model thinking continued to remind Defense that it was safest to describe as purely applied any research that bore on the Defense Department's mission.

The Atomic Energy Commission achieved an amalgam of pure and mission-oriented research at a far more wholesale level of allocation in the early postwar years. Wanting to enlist the help of the remarkable generation of physicists who split the atom and built the bomb, and coping with the dangers of a divided world, the AEC encouraged their involvement by its willingness to support, at increasingly generous levels, the theoretical and experimental research in high-energy physics that laid the foundation of the Standard Model of the elementary particles of the physical universe. The implicit bargain in these arrangements was symbolized by the presence of both weapons-directed and curiosity-directed research in the Argonne, Brookhaven, Livermore, Los Alamos, and other national laboratories,[22] although this link weakened when it was clear that the research in high-energy physics was as pure as any in the postwar era. By the time Enrico Fermi's generation gave way to the generation of Sydney Drell, and DOE supplanted the Atomic Energy Commission as sponsoring agency, high-energy physics was virtually a freestanding program of pure research, almost anomalously housed within a Department of Energy.

The distorting effects of the postwar paradigm are clearer still in the way that the failure to recognize the fundamental character of use-inspired science in Pasteur's quadrant has inhibited the emergence of basic research that is germane to agency missions. A notable example is furnished by the drive for fusion energy and the development of plasma science, an example that touches important issues of current science policy.

Although the physics of nuclear fusion as the source of the radiant energy of the sun and stars was spelled out earlier in the century, it was the work on thermonuclear weapons after World War II that created the dream of reigniting the sun on earth as a source of virtually limitless power with far less environmental hazard than fission power entailed. From the start it was clear that this goal would require an immense engineering effort, and it was also clear that it would require fundamental advances in plasma science. Subsequently, each of these technological and scientific trajectories has profoundly influenced the other, and their interactive relationship is recognized by a recent report issued by the National Research Council, which notes that "plasma science is central to the development of fusion as a clean, renewable energy source"[23] and that "magnetic confinement fusion continues to be the largest driver for the intellectual development of plasma science."[24] Well it should, in view of how clearly the realization of this technological dream will require a fundamental understanding of such things as turbulence in plasmas and the turbulent transport of particles and energy.

The several-decade quest of this fundamental knowledge might have gone quite differently if the policy and scientific communities had held more realistic views of the relationship between the advances of basic science and of technology. In this case, the *policy* community might have seen this investment in scientific knowledge as a natural corollary of achieving an important technological goal. And the *scientific* community might have seen the drive for fusion energy as a striking opportunity to develop fundamental knowledge of matter in the plasma state. As it was, the vision of both communities has been clouded by the tendency to see the world of research as divided into the pure and purely applied that is associated with the postwar paradigm.

The *policy* community's perspective on the drive for fusion power reveals how effectively the postwar paradigm has impeded the upward diffusion of research in the mission agencies from Edison's to Pasteur's quadrant. Once the technological goal of generating commercially profitable fusion energy was defined, agency program managers and their funders tended to style the effort as a huge venture in applied research and engineering, akin to the Apollo space program that put a man on the moon with off-the-shelf science. Like the Apollo program, the drive for commercially profitable fusion energy was soon geared to the construction and operation of ever larger and more powerful machines. In the course of scaling up, the leadership role of the scientists who first articulated the dream of rekindling the power of the sun was increasingly shared with engineers and managers who could design and operate these machines within the constraints of time and budget. Appropriations requests were keyed to these technological milestones, and the need to extract a heavy flow of public funding created incentives for minimizing the uncertainties ahead, including those surrounding the fundamental scientific knowledge that igniting a fusion reactor would require. These uncertainties have in particular been understated by the Energy Department's Office of Fusion Energy, the lead agency in the drive for fusion power. And as budgets became tighter, Congress reinforced the incentives for minimizing the scientific uncertainties surrounding fusion energy by focusing on tokamak, a single energy concept, as the most promising, to the detriment of research on alternative concepts.[25]

But it could equally be said that the *scientific* community's perspective on the drive for fusion power reveals how the postwar paradigm has also impeded the upward diffusion of mission-agency research from Edison's to Pasteur's quadrant. Once the technological goal of generating commercially profitable fusion energy had been defined, much of the physical science community too saw this venture as a huge program in applied research and engineering rather than as a major opportunity to explore the fundamental phenomena of plasma physics.

Hence, the complementary perspectives on the drive toward fusion power held by the policy and scientific communities, each deeply influenced by the postwar paradigm, have limited the fundamental Pasteur's quadrant research that might have flowed from this drive. After several decades of unparalleled support for the associated engineering and development, the NRC panel on plasma science could observe that

> we have no first-principles understanding of turbulence in *any* plasma, and understanding such turbulent behavior is perhaps the key unsolved problem in plasma physics.[26]

This conclusion was reached by the panel despite the fact that virtually all magnetically confined plasmas are profoundly influenced by this turbulence.[27] The panel went on to say that

> because turbulence and turbulent transport are not understood in any plasma, careful experimentation in flexible, small experiments is likely to make significant contributions to testing existing theoretical predictions and to guide further theoretical work in this important area. Given the fundamental lack of understanding and the important practical consequences that would derive from a deeper understanding of turbulence and turbulent transport, a sustained program of both theoretical and experimental research is extremely important.[28]

The quest for this fundamental knowledge can benefit from "small science" as well as from inquiries mounted with large machines. Just as it is important for the drive toward fusion energy, which requires very large installations, not to crowd out funding for fundamental research in much smaller laboratories, so it is important for the large installations not to be wholly seen as handmaidens to a technological goal.

Comprehensive Investment Centered on Pasteur's Quadrant

An extraordinary match between societal needs and research seeds accounts for the spectacular growth of the National Institutes of Health in the postwar decades. The needs were compelling, in view of the priority given by the policy community and public to the goal of health. This priority is shown by the steadily rising fraction of gross national product directed to health care, as well as by the soaring government expenditures for biomedical studies.[29] In the words of John Sherman, "Every public opinion poll since World War II touching on the subject has reflected a strong willingness to see our federal tax monies used for the support of medically related research,"[30] and for several decades there was a W. S. Gilbert quality to the way the president's handsome requests on NIH's behalf were regularly topped by the appropriations voted by Congress. The seeds were also favorable, since the fields of molecular biology, biochemistry, and allied sciences were capable of an explosion of biomedical discovery and unparalleled advances in the understanding of disease and its remedies. An important additional background factor may have been the presence of a professional group—the country's doctors—schooled in the biomedical sciences to at least an elementary level. This confluence made NIH the most successful channel of support for use-inspired basic science in the postwar era, and this development was spectacularly validated when it was realized how applicable the knowledge gained by fundamental research in molecular biology would prove to be.

From the moment NIH offered to assume responsibility for the contracts of the Committee on Medical Research of the Office of Scientific Research and Development, as OSRD went out of business at the end of World War II, NIH's leadership skillfully shaped a distinctive research role for the institutes. The years of James Shannon's leadership from the mid-1950s to the late 1960s were marked by the decision to achieve NIH's extraordinary growth primarily through the program of peer-reviewed grants rather than work within NIH's own laboratories, a decision that helped unleash the postwar revolution in biological science in the universities. The human significance of biomedical knowledge since the time of the Hippocratics placed the center of gravity of this research squarely in Pasteur's quadrant. But the evolving strategy of NIH coupled this with substantial investments in pure research and purely applied research as well. What emerged was a comprehensive strategy, unique in America's experience, of research investments that included all three patterns of research goals but was clearly centered on use-inspired basic science, an institutional strategy that has led at times to a kind of schizophrenia among both NIH staff and principal investigators. In policy circles, they are apt to emphasize their Pasteur's quadrant role, whereas in academic research circles, where the ideal of pure inquiry still burns brightly, they are apt to emphasize their Bohr's quadrant credentials.

The evolution of federal support in the postwar decades has therefore channeled funds to use-inspired basic research by varied channels involving agencies that invoke varied criteria in their wholesaling of such support. The scientific R&D agencies heavily rely on categories that are defined in terms of disciplines or fields.

By contrast, the support for use-inspired basic science by the R&D-intensive mission agencies flows through channels that are problem or mission defined. Thus the Department of Energy establishes the Office of Basic Energy Sciences. The Agriculture Department proposes research initiatives on natural resources and the environment; plants; animals; nutrition, food safety, and health; processes and new products; and markets, trade, and rural development.

The outlook of these two groups of agencies is therefore fundamentally different. The significance of this difference may be captured by a metaphor of hammer and nails borrowed from the philosopher of science, Abraham Kaplan, and his *"law of the instrument."*[31] In this homely figure, the deepening understanding sought by the scientific R&D agencies may be likened to making a better hammer; the problems that this understanding may help solve, likened to the nails that such a hammer could more effectively drive home. To be sure, these agencies know quite a lot about the nails that need hammering, as NIH knows a great deal about the diseases that might be cured by the gain in scientific understanding from research it sponsors; in this sense, a good deal of this research is use inspired. But these agencies are first of all in the business of improving hammers, of raising scientific understanding to a higher level from its current disciplinary base.

The mission agencies are, by contrast, in the business of driving in nails, and ordinarily look for a better hammer when they need to drive such a nail home. These mission-related objectives (the nails) provide a distinctive perspective for the basic scientists whose work they fund, a perspective that leads at times to strikingly original research. The value of this fresh angle of vision extends the list of "standard" advantages of a plural system of research support over a monolithic system—the greater possibility of multiple, independent reviews of novel research ideas and the greater continuity of support that plural sources offer academic and other organizations performing research.[32]

The country has almost certainly been well served by the pluralism of these arrangements and has reason to be skeptical that an overarching ministry of research, adapted to American circumstances as a Department of Science, would preserve this value. This observation is important enough to deserve restating:

The flows of support through the multiple channels of the scientific and R&D-intensive mission agencies are a source of creativity in building agendas of use-inspired basic research, a source that would be put at risk by the formation of an overarching Department of Science.

If the evolution of federal support in the postwar decades has managed to fund a use-inspired basic science that is largely concealed by the postwar paradigm, one ought not to miss the degree to which this paradigm has impeded the flow of support to basic research of this kind. It has in fact hampered the development of a broader research mission by the National Science Foundation and the development of stronger programs in basic science by the R&D-intensive mission agencies

Evaluating the NIH Model

The distortions created by the postwar paradigm that impeded the support of fundamental yet use-inspired research by NSF and the mission agencies have been far less troublesome in the case of the National Institutes of Health, an agency that grasped the importance of use-inspired basic research from the inception of its extramural grants program after World War II. A comprehensive strategy of research support anchored on Pasteur's quadrant positioned NIH to resist far more effectively the pressures that the postwar paradigm brought to bear on NSF and the R&D-intensive mission agencies.

This benefit of a research strategy that granted full, if implicit, recognition to the science of Pasteur's quadrant was shown in the late 1960s and early 1970s by NIH's success in coping with the rising enthusiasm of the policy community for putting new knowledge to work. A shudder went through the biomedical community when Lyndon B. Johnson chose NIH's Bethesda campus as the site of an address in which he implied that biomedical science had amassed enough new knowledge and that this knowledge should now be taken from the shelf and turned into cures for the diseases that still threatened the nation's health. This apparent case for applied research and development was echoed by the War on Cancer declared in 1971 by Johnson's successor, Richard M. Nixon. Parts of the biomedical research community understood the War on Cancer to be a call for fairly narrowly targeted applied work.

What kept these calls for applied work from having a far greater impact on NIH's priorities was the belief of the biomedical community that work of fundamental scientific significance could be inspired by societal need. NIH may have felt it had been provided with more money for cancer research than it could in the short run usefully spend. But it did not believe that this windfall appropriation had to be deployed on narrowly applied studies. On the contrary, it knew how to dedicate much of the added funding to Pasteur's quadrant research that advanced fundamental scientific understanding as it helped with the detection, treatment, and cure of human cancers. An example cited by Daniel J. Kevles is the fact that the "war-on-cancer money paid for basic research into the mechanisms that transform healthy cells into malignant ones, and so sustained the work that led J. Michael Bishop and Harold Varmus . . . to their Nobel Prize—winning discovery of oncogenes."[33]

This lesson from the history of the National Institutes of Health underscores the fact that the country has not adequately exploited the varied institutional experience it has gained in the wholesaling of support for use-inspired basic science. There is no reason to believe that the enormously successful model pioneered by NIH must be confined to biomedical research. Important factors favored its success there. The goal of better health commanded universal support. The disciplinary range of the biomedical sciences was relatively limited. These sciences stood at the edge of discoveries of profound importance for the country's unmet health needs. And much of the needed research was "small science" that could be done by individual investigators and teams of researchers, supported by competitive, peer-reviewed grants. Yet these factors ought not to be overstated. The debates over abortion and animal rights remind us of potential conflicts surrounding health care, and NIH has avoided other sources of conflict, such as those which might have been generated by flourishing programs of research on the health risks of tobacco use or the economics of health care delivery.

Plainly, the NIH model could be repeated only sparingly. The resources that built up the National Institutes of Health were invested over several decades of flush federal budgets and rising incomes. Under the budgetary constraints that will be in force for the rest of the century, any replication would require reallocating funds. It is nonetheless unreasonable to think that only biomedical research lends itself to the virtues of the NIH model of a scientific agency focused on an area of recognized societal need that is able to enlist basic science of the highest caliber from a range of disciplines to develop a fundamental understanding of the phenomena that underlie the problem area, while it also sponsors some pure research and some purely applied research.

The environment is a problem area in which the NIH model might be replicated. Over several decades a growing multidisciplinary community of scientists has laid the groundwork for such an experiment by the understanding they have gained of the global and more regional systems that shape the human environment. Physicists, chemists, oceanographers, geologists, meteorologists, ecologists, and scientists from many other fields have made an impressive start on understanding such basic processes as the "grand cycles" by which carbon, nitrogen, sulfur, and phosphorus move from the soil, the atomosphere, and the oceans to living plants and animals and back again. Understanding these cycles in the millions of centuries prior to the appearance of the human species provides a backdrop for understanding the subsequent human influence on these four biochemical building blocks for all forms of life, an influence that has sharply escalated in recent centuries. D. Allan Bromley, the distinguished physicist who served as George Bush's science adviser, frequently cited the estimate that more than 200 billion tons of carbon have been injected into the atmosphere since 1850.[34]

Although it is an extraordinarily difficult scientific task to separate human influence from natural variation in these complex systems, this research has charted the impact of human technology on most of the basic biochemical cycles of the natural world, an impact leading to consequences as disparate as acid rain and global warming. This research has enjoyed increasing success in spelling out the nature of such fundamental processes as the exchanges between the atmosphere and the world's oceans and the interface between the inorganic world and biological forms as varied as the tropical forests and sea-borne algae. But much more needs to be learned to provide a sure understanding of how we can preserve the essential balances in the human environment in a new century of rapid technological change.

Reproducing the NIH model in the environmental area would need to be justified by judgments of research promise at the wholesale level. It is unlikely that replicating the NIH model would lead to advances in scientific understanding comparable to those achieved by the biomedical sciences in the postwar decades, although the creation of such a funding channel would encourage important breakthroughs. The National Academy of Sciences has been actively tracking the potential of research on global change, and the National Science Foundation has mounted small but vigorous programs of fundamental science on the environment. Perhaps the most revealing indicator of scientific promise is the remarkable number of research universities that have mounted environmental initiatives focused on both teaching and research. The sense of research opportunity shared across a number of academic institutions has indeed already spawned an effort to create a National Institute for the Environment.[35] It is quite possible that the emergence of an important new channel of funding of rigorous, peer-reviewed projects would further unite the research strengths of a number of scientific and engineering fields that bear on the environment.

Equally critical judgments of societal need would be required to justify the NIH model in the environmental area. As the importance of protecting the environment burst on the nation's consciousness in the late 1960s and early 1970s, this need may have enjoyed as strong a consensus as the importance of improved health, and the opinion polls have continued to record overwhelming majorities of the American people as supporting clean air and clean water in subsequent decades. Nonetheless, a more polarizing belief in the trade-off between environmental protection and economic opportunity has characterized the debate over environmental policy in recent years. Indeed, the appeal of an NIH for the environment is diminished in some political quarters by the fear that deeper scientific understanding would do little more than give future government regulators additional tools to constrain industry. But an NIH-type investment in the environmental area would attract broader political support if a deeper scientific understanding was expected to yield an economic return for the private sector —by allowing firms to experience the greater profitability of sustainable development or to exploit some of the resulting technologies, as the pharmaceutical firms have profited from the fundamental advances in biomedical science. On the latter point, the United States might well learn from the Japanese, who have remarkably clear ideas as to how an investment in basic environmental science may help Japanese firms to sell the rest of the world the technologies that will be needed to clean up the global environment in the next century.

The environment is not the only area in which the match of research promise with societal need might justify repeating the NIH model. So important is the intense interplay of fundamental science and technological development in a cluster of fields surrounding condensed-matter physics and materials science that this area too might supply an opportunity for an NIH-type channel for the wholesaling of research support that could enlist the interest of the basic research community and protect the scientific integrity of decisions at the project level. In any case, the conclusion to which these observations naturally lead can be stated in general terms:

> In the evolving arrangements for funding use-inspired basic research, attention should be given the possibility of replicating the National Institutes of Health model in other areas where research promise is felt to match societal need, with the field of environmental science a leading candidate.

At the retail level, choices among alternative projects of use-inspired basic research should be largely in the hands of those who can judge scientific promise, constrained by wholesale decisions as to the societal needs a given program of research is to address. By contrast, there is at the *wholesale* level of research funding much less asymmetry between the ability of those with scientific expertise to judge the social value of alternative programs of research and the ability of nonscientists to judge the scientific promise of these programs. Where these decisions are confined to a single scientific field, something akin to the central limit theorem of probability theory may make it easier for scientists well versed in the field to gauge the promise of significant progress by a large number of projects than to predict the success or failure of particular projects. But this advantage rapidly melts away when the promise of alternative programs of research must be judged across fields of widely varied conceptual and empirical content.

Yet these constraints by no means remove the need to bring the best scientific judgment to bear on wholesale decisions of support for use-inspired basic science. At this level too it is critically important to join informed estimates of scientific promise with judgments of societal need, lest those in political authority fund research bearing on their social goals that is as likely to affect the course of affairs as King Canute's commands to halt the advancing tide. The annals of science policy in the postwar decades chronicle a sustained effort to institutionalize the flows of scientific information to the president and Congress, the twin centers of democratic authority in American national government.

Linking Scientific Judgment with Political Authority

The effort to bring together informed judgments of scientific promise and of national need date from the earliest years of the federal investment in research during World War II. Indeed, the blending of scientific expertise and political authority in the wartime presidency of Franklin D. Roosevelt set a standard that was difficult for Roosevelt's postwar successors to meet. The role of science adviser almost disappeared from the White House in the earliest postwar years. For science policy advice, Harry S. Truman primarily relied on John R. Steelman, an economist and sociologist by training, who chaired the President's Scientific Research Board and supervised the compilation of the five-volume Steelman Report.[36] The scientists brought into the Executive Office in Truman's presidency served only as advisers to the director of the Office of Defense Mobilization, without direct access to the president.[37] It took the storm created by the launching of *Sputnik* in

Eisenhower's second term to revive a relationship closer to the one that F. D. Roosevelt and Vannevar Bush had pioneered. In response to this Soviet challenge, Eisenhower appointed James R. Killian, president of the Massachusetts Institute of Technology, as special assistant to the president for science and technology, and Killian assembled a group of volunteer colleagues to serve as the President's Science Advisory Committee (PSAC), also with direct access to the president.

These arrangements were strengthened by John F. Kennedy. His special assistant for science and technology, Jerome B. Wiesner, became an explicit part of the Executive Office as director of an Office of Science and Technology and played an important role in the soaring expenditures on space and defense R&D as the confrontation with the Soviets entered its tensest stage. But this advisory relationship suffered as the Vietnam War darkened the feelings toward academic science held by Kennedy's successors, Johnson and Nixon. Indeed, Nixon was so angered when members of PSAC went public with their opposition to his Anti-Ballistic Missile, Super-Sonic Transport, and Vietnam policies that he dissolved the committee and dismissed Edward E. David Jr., his science adviser, declaring that he would henceforth take counsel on science and technology policy from the director of the National Science Foundation.

The recovery from this low point for science advice in the White House proceeded in several stages. President Gerald Ford reinstated the role of science adviser, with a willing Congress providing statutory authority for a White House Office of Science and Technology Policy (OSTP), directed by the science adviser, and for a Federal Coordinating Council for Science and Technology—later changed to Science, *Engineering*, and Technology (FCCSET)—chaired by the science adviser. Guy Stever, the director of the National Science Foundation who doubled as science adviser after Nixon abolished the White House post, left the NSF to devote his energies to these arrangements.

Frank Press, who succeeded Stever as science adviser in the Carter years, played a critical role in shaping the priorities for scientific research during the administration of a nuclear engineer turned president. Press formed a close entente between OSTP and the Office of Management and Budget and pioneered the concept of a presidential R&D 'initiative' cross-cutting a number of departments and agencies.[38] An agreement between Press and Bert Lance, Carter's first director of OMB, which was continued by Lance's successor, James McIntyre, allowed senior OSTP staff to play a role in R&D decisions at each step in the construction of the executive budget right up to the director's review. But the revival of a committee of advisers from the scientific community fell victim to Jimmy Carter's desire to reduce White House staff and to mold the Executive Office to the president's decision needs.

Although Ronald Reagan believed that basic science was a prime element of the country's strength in confronting the Soviets, the Reagan presidency took at least a short step backward in the role of science advice in the White House. George Keyworth, the first of Reagan's directors of OSTP, was a late recruit with limited stature in the scientific community and was felt to be too much within the orbit of Edwin Meese, Reagan's principal policy adviser. Despite these reviews, Keyworth and his successor, William Graham, accomplished a good deal, including launching the ultimately star-crossed Superconducting Supercollider.

The revival of science advice in the White House continued during the presidency of George Bush, who for the first time made his science adviser one of the assistants to the president.[39] Bush also took advantage of the previously unused statutory authority to revive and meet directly with an advisory committee of scientists—renamed the President's Committee of Advisors on Science and Technology (PCAST). Beyond these steps, he agreed to fill all of the senior staff positions provided by the statute creating OSTP. This added staff strength permitted the Bush administration to mount important initiatives in science and technology, most notably in the fields of environmental science, high-performance computing, advanced materials, and biotechnology.[40]

Science advice in the White House entered a further stage in the Clinton administration, which considerably elaborated the organization of this function as it defined and implemented its priorities in science and technology, including several inherited from the Bush administration.[41] Clinton, a president with a strong technology agenda, created a White House National Science and Technology Council (NSTC), a cabinet-level group chaired by the president himself that is formally coordinated with the National

Security and National Economic Policy Councils. The dual post of the assistant to the president for science and technology and director of the Office of Science and Technology Policy was filled by John Gibbons, an experienced former director of the congressional Office of Technology Assessment, who in his Capitol Hill years had become close to the new, technology-minded vice president. One of nine committees of the NSTC was charged with overseeing the federal role in basic science and was jointly chaired by OSTP's associate director for science and by the directors of the National Science Foundation and National Institutes of Health. Like its predecessor, the Clinton administration impaneled a PCAST committee of scientific advisers and asked its members and others from the scientific community to undertake such ad hoc policy studies as the previously cited review of fusion energy. The experience with this apparatus in a presidency as favorable to science and technology as any in the postwar period suggests the limits of what can be done in the eye of the governmental hurricane to bring deeply informed judgments of scientific promise to bear on wholesaling decisions on research support. Both the time and preferences of the president were mortgaged to a vast array of other commitments. Likewise, the officials who sat on the NSTC and its committees had departments and agencies to run and turf to protect. Their staffs, drawn from OSTP and elsewhere in the executive branch, were stretched equally tight by short-term deadlines and fires that needed to be put out. Although some of the scientists staffing this apparatus brought agendas that reflected their own prior involvements in research, on most issues they served as brokers mustering scientific and political information for those in the White House and OMB who were reaching decisions on the wholesaling of support for research and development.

The effort to institutionalize the flow of science advice has had a still more checkered history in Congress, the other major institution with legitimate authority over the wholesaling of support for scientific research in our democratic system. In the first postwar years Congress sought technical information to deal with science and technology policy issues by the classic means of elaborating its committee structure. The Joint Senate-House Committee on Atomic Energy, created by the Atomic Energy Act of 1946, oversaw all aspects of the work of the newly established Atomic Energy

Commission, including appropriations, from 1947 to 1977, when the committee was dismantled and the regulation of nuclear energy divorced from the operating responsibilities that the Department of Energy inherited from the AEC. In the wake of *Sputnik* the Senate created a Select Committee on Space and Aeronautics that in 1958 was transformed into a Standing Committee on the Aeronautical and Space Sciences. In 1977 the Senate Commerce Committee became the Committee on Commerce, Science, and Transportation, with responsibility for authorizing and overseeing the government's science and technology programs. In 1958, in the scientific build-up following *Sputnik*, the House of Representatives followed suit by creating a Standing Committee on Science and Astronautics. In 1974 this became the Committee on Science and Technology; in 1987, the Committee on Science, Space, and Technology; and in 1995, simply the Committee on Science, as a reflection of the skepticism about the government's technology programs harbored by the new Republican leadership. In line with the deep congressional tradition of separating appropriations from authorization and oversight, the funding of scientific programs remained the preserve of the appropriations committees and sub-committees of the House and Senate, apart from the three-decade exception of the Atomic Energy Committee.

The hearings of these committees were an important channel for scientific information to reach the government, and their proceedings and reports are a distinguished archive of American science and technology policy. The formation of these committees and rapid expansion of congressional staffs in the postwar decades brought into Congress's service a number of scientists and engineers who could play important interpretive roles for their members and committees.[42] Two support agencies of Congress also acquired increasing staff strength in science and technology policy to respond to requests by the authorizing and appropriating committees of Congress and by individual members. At the General Accounting Office (GAO) successive comptrollers general welcomed the need to provide congressional committees with advice on science and technology programs as part of their efforts to extend GAO's audits to include program performance as well as financial management. Elmer B. Staats, the comptroller general who pioneered this change, had a deep background in science and technology policy from his years at the Bureau of the Budget. At the Library of Congress, the Legislative Reference Service was upgraded in the postwar decades, becoming in 1972 the Congressional Research Service (CRS). From the outset CRS created a Science Policy Division and an Environmental and Natural Resources Division to respond to requests from individual senators and representatives for information on science and technology policy issues.

Congress took a far more ambitious step to provide itself with advice on science and technology policy when it created the Office of Technology Assessment (OTA) in 1972. The title of this agency reflected the concern of many of its congressional sponsors over threats to the environment posed by new and rapidly evolving technologies in a period when the environmental movement was at flood. But the reports issued by OTA covered a broad terrain and embraced a number of scientific questions as well. Moreover, these reports, largely contracted out to specialists in the field, were often of a scope and depth that reached well beyond the policy and program analysis more typical of the reports issued by GAO and CRS.

OTA did not have an easy passage, for reasons that illuminate the tensions between political interest and scientific truth. As it successfully tried to serve each of these masters, the agency was twice forced to change course in its early years. Its first director was a former congressman, Emilio Daddario, who five years earlier had shepherded the changes in NSF's charter through Congress. He and a Technology Assessment Board drawn from the members of the House and Senate administered OTA as they might have run a congressional committee. Within two years Daddario fell before a fusillade of criticism from those who expected the agency's reports to be neutral and analytical. So fully did OTA's second director, Russell Peterson, take account of the critics' complaints that he defined a strategic agenda that largely ignored the intent of OTA's organic act to create an agency to help Congress in the legislative assessment of matters pending. Peterson's term came to an even more abrupt end in a shower of complaints that OTA was proving irrelevant to the work of Congress. Its third director, John Gibbons, later the science adviser to President Clinton and director of the White House Office of Science and Technology Policy, sought to steer a course between these two shoals by undertaking studies that were analytically objective but enjoyed bipartisan support and were keyed to the legislative agenda of Congress.

Although OTA earned substantial respect in the twelve years of Gibbons's tenure, it never entirely surmounted the tensions inherent between the needs of policy research and of legislative decision. These tensions were twofold. One had to do with time. The time needed by policy research to reach an analytically sound result may be poorly matched to the rhythms of the legislative process, which may require a decision on the basis of the current balance of political forces and what is by then known. This mismatch continued to frustrate members of Congress even when OTA sought to limit this dissonance by undertaking short-term studies keyed to the legislative timetable. The other tension is between analytical conclusions and political interests. However sensitively a study may take account of conflicting interests, its conclusions can seem injurious to some of the stakeholders. This complaint acquired a partisan cast when the shift of OTA's director and several of his senior aides to the Clinton White House seemed to confirm the agency's liberal Democratic orientation during his long tenure.[43]

OTA proved vulnerable on both these grounds when the Republican leadership brought to power by the elections of 1994 set out to cut Congress's own budget prior to balancing the budget of the government as a whole. OTA was an easy mark for the budget cutters as an agency with several hundred employees who could be eliminated at a stroke. But its 1995 death also reflected the inherent difficulty of building a constituency for an agency dedicated to science and technology policy research in an institution with exceedingly brief time horizons and highly political tests of truth.

Despite the vast differences in the problem of advising a unitary executive and a plural legislature, several closely related lessons are to be learned from the periodic travails of science advice in each of these settings:

The Primacy of Political Considerations. The fragility of these channels of advice shows how easily scientific judgments can be overwhelmed by political considerations when the stakes are high for those in political authority. The stronger the political agenda of elective leaders the less open they will be to research-based advice. It is unreasonable to suppose that a rationalist, problem-solving view of the role of science will prevail when those in authority believe that the results of use-inspired research may be unwelcome or threatening.[44] It is not surprising that a political agenda should color the view of scientific advice that was held by a Lyndon Johnson embattled in Vietnam or by a new congressional leadership determined to implement its Contract with America.

Discounts of the Future. It is understandable in an elective democracy for the policy horizons of those in elective office to coincide with its two-, four-, and six-year election cycles. Democratic politicians are by no means the only leaders in our society whose discounts of the future are thought to be excessively deep. It has become almost a commonplace to criticize corporate leaders for focusing too much on their firms' performance in the next quarter and too little on returns in the longer run. Indeed, many corporations began their downsizing of central research laboratories in the high-inflation years following the oil shocks of the 1970s, years in which the deepest discounts were applied to the future.

From his observation of successive presidencies, Staats has called the reconciliation of the president's short-term need to marshal support in Congress and the public and his concern to serve the nation's long-term interest one of the deepest tensions of the science advisory process.

> Long-term analysis . . . requires extraordinary discipline to prevent resources assigned to strategic planning from being absorbed into short-term "firefighting" work. . . . To be effective, strategic planning requires insulation without isolation.[45]

Staats's concern led him to emphasize the importance of the President's Science Advisory Committee (PSAC), reborn in the Bush administration as the President's Committee of Advisors on Science and Technology (PCAST), as a source of long-term thinking.

Focus on Decision Needs. From what has been said it follows that those who hold political authority will want policy advice that is relevant to the matters with which they must deal. Presidents and Congresses will want their science policy advisers to march to the rhythms of the policy process and to tailor their structures for policy advice to their decisionmaking needs. Among postwar presidents, Jimmy Carter most clearly articulated this principle as he devolved the reporting duties of the White House Office of Science and Technology Policy to the National Science Foundation and declined to reconstitute a PSAC committee. His criterion, that "the Executive Office of the President exists to serve the President and should be structured to meet his needs,"[46] has guided the actions of a number of chief executives as they have redesigned the science advisory apparatus of the White House.

This pervasive need makes it highly unlikely that Congress or the Executive Office will provide a setting for a unit that is equipped to mount a continuing and deeply informed assessment of the scientific promise of the alternative programs of use-inspired basic research that might be funded by the national government. In our open policy system, the role of the science advisory structures at both ends of Pennsylvania Avenue is inevitably one of marshaling such assessments from other quarters. This is a classic role of the science and technology oversight committees and subcommittees of the House and Senate. In the Bush and Clinton administrations, the PCAST committee of presidential advisers has been increasingly effective in providing high-level reviews of such things as megascience projects and the future of fusion energy research mounted by panels that include both PCAST members and specialists recruited for a particular review.

In the American system, an important role in assessing the research promise of alternative fields is played by the congressionally chartered National Academy of Science, National Academy of Engineering, Institute of Medicine, and their operating arm, the National Research Council (NRC). With funding from Congress, the National Science Foundation, and other governmental sources the NRC has brought the specialized knowledge of the scientific community to bear on a remarkable range of science policy questions, including the promise of major advances in a broad spectrum of scientific fields and subfields. An important, continuing role is also played by the famously successful American Association for the Advancement of Science, publisher of *Science* magazine, where the continuing program of analytic work on the science and technology budget deserves its enviable reputation. The American system also depends on the science policy work undertaken by a remarkable array of academic and freestanding policy research units.

It is difficult not to feel, however, that there is a need for a sustained and well-institutionalized effort within the national government itself to canvass the strength of the scientific fields that can be brought to bear on national needs. To raise this question is to echo Bush's call for a National Research Foundation a half century ago. It is also to raise the question of whether such a role could be played by the National Science Foundation and National Science Board, the institutional heirs to Bush's dream. The National Science Board was charged with overseeing all government support for science by the act creating NSF in 1950. But Waterman, the Foundation's first director, was recruited from the Office of Naval Research, one of the several units of the government that long since had moved into the vacuum created by the disappearance of the Office of Scientific Research and Development at the end of the war. He had little taste for the trench warfare with the departments and agencies that would have been needed, with the Bureau of the Budget's backing, to assert this authority, and later members of the National Science Board settled into the common wisdom that the time for such a role had passed.

This question ought to be reopened in a contemporary form. Indeed, one of the benefits of the Foundation's playing this role would be to make clear to Congress that the key to turning the power of basic science to national needs is not to ratchet upward from 50 to 60 percent the proportion of NSF's budget that supports "strategic" research but to build agendas of use-inspired basic science, funded by agencies across the government, that bear on the nation's needs. Just as the Foundation can play a vitally important role in helping to safeguard the scientific integrity of choices among alternative projects at the retail level, it could play an important role in promoting the *scientific* integrity of choices among alternative programs of use-inspired basic research at the wholesale level. It is unlikely that the open American policy system will lend itself to an analog of the Japanese system of research foresight. But the need to bring deeply informed judgments about research promise to bear on the allocation of funds to research focused on national needs is as clear today as it was a half century ago, even if the compact between science and government is redrawn on premises very different from the belief in the return on research performed without thought of practical ends.

NOTES

Chapter One

1. See chapters 2 and 4 for an account of the attacks on the postwar consensus in the period of the Vietnam War and its aftermath.

2. Vannevar Bush, *Science—The Endless Frontier: A Report to the President on a Program for Postwar Scientific Research* (Washington: National Science Foundation, reprinted 1990).

3. See, for example, the British government's White Paper, *Realising Our Potential: A Strategy for Science, Engineering and Technology,* Cm 2250 (London: Her Majesty's Stationery Office, 1993), p. 15, which declares that "basic research is, by definition, research without a specific end in view."

4. Bush, *Science*, p. 18.

5. Ibid., p. 83. Emphasis in original.

6. Ibid., p. 19.

7. Ibid., p. 19. Emphasis in original.

8. Harvey Brooks, "Basic and Applied Research," in *Categories of Scientific Research*, papers presented at 1979 National Science Foundation seminar, Washington, pp. 14–18.

9. OECD Directorate for Scientific Affairs, *The Measurement of Scientific and Technological Activities: Proposed Standard Practice for Surveys of Research and Experimental Development: Frascati Manual 1993* (Paris: Organization for Economic Cooperation and Development, 1994), p. 29.

10. James Bryant Conant, ed., *Case 6: Pasteur's Study of Fermentation* (Harvard University Press, 1952), p. 9.

11. Bush, *Science*, p. 83.

12. This view was frequently echoed by members of the scientific community when they were called upon to explain the nature of research. Glenn T. Seaborg informed a committee of Congress that "the motivating force [in basic research] is not utilitarian goals, but a search for a deeper understanding of the universe and of the phenomena within it." [see *Federal Research and Development Programs*, Hearings before the House Select Committee on Government Research, 88 Cong. 1 sess. (Government Printing Office, 1964), pt. 1, p. 66]; Leland Haworth, that basic research "seeks an understanding of the laws of nature without regard to the ultimate applicability of the results" (see *Federal Research and Development Programs*, Hearings, pt. 1, p. 6); Edward Teller, that such research "is a game, is play, led by curiosity, by taste, style, judgment, intangibles" (see *Government and Science*, Hearings before the Subcommittee on Science, Research, and Development of the House Committee on Science and Astronautics, 88 Cong. 1 sess. (GPO, 1964), p. 115).

13. *Second Annual Report of the National Science Foundation Fiscal Year 1952* (GPO, 1952), pp. 11–12. Emphasis added.

14. "Paradigm" as used here has only a loose kinship with the term as employed by Thomas S. Kuhn. I do not invest this way of looking at science itself with all of the properties of scientific paradigms in Kuhn's analysis. See Thomas S. Kuhn, *The Structure of Scientific Revolutions*, 2d ed. (University of Chicago Press, 1970).

15. Parliament of the Commonwealth of Australia, *Report of the Committee on Australian Universities*, September 1957 (Canberra: Commonwealth Government Printer, 1958), p. 9.

16. See in particular Merton's 1957 presidential address to the American Sociological Society, reprinted as chapter 14, "Priorities in Scientific Discovery," in Robert K. Merton, ed., *The Sociology of Science: Theoretical and Empirical Investigations* (University of Chicago Press, 1973), pp. 286–324.

17. The complexity of the relationship between these goals in Pasteur's work is one of the themes of the remarkable study by Gerald L. Geison, *The Private Science of Louis Pasteur* (Princeton University Press, 1995). Working from previously inaccessible materials, Geison shows in particular what an astonishing portion of Pasteur's understanding of microorganisms was in place when he began his investigations in Lille. In this sense he was eager for—and may indeed have helped inspire—the appeal for help from those who manufactured alcohol from beets. This mix of discovery and practical triumph characterized the whole subsequent course of Pasteur's career, as he worked out the details of his germ theory of fermentation and disease. As Geison is at pains to note, the gap between the Pasteur legend and the working scientist that is revealed by his notebooks does nothing to diminish the stature of his scientific work. Indeed, Geison gives fresh evidence across a century's time of the extraordinary power of Pasteur's scientific insight.

18. See Crosbie Smith and M. Norton Wise, *Energy and Empire: A Biographical Study of Lord Kelvin* (Cambridge University Press, 1989).

19. Charles C. Gillispie, *The Professionalization of Science*, The Neesima Lectures (Kyoto: Doshisha University Press, 1983), p. 12. See also John W. Servos, "The Industrial Relations of Science: Chemical Engineering at MIT, 1900–1939," *Isis*, vol. 71, no. 259 (1980), pp. 531–49. A second group of research advances that Gillispie saw as helping to make American science world class—Morgan's research on genetics—was also deeply influenced by considerations of use.

20. The views of J. Robert Oppenheimer and Henry DeWolf Smyth cited here were almost certainly influenced by their belief that the most fundamental research seeks to discover new laws of wide generality rather than to explain new phenomena in terms of known principles; that is, that fundamental research is, in Weisskopf's terms, "intensive" rather than "extensive."

21. *First Annual Report of the National Science Foundation: 1950–51* (GPO, 1951), p. 10.

22. Ibid., p. 10.

23. Sir Arthur Lewis, "The Slowing Down of the Engine of Growth," Nobel Memorial Lecture, December 8, 1979, in *Les Prix Nobel 1979: Nobel Prizes, Presentations, Biographies, and Lectures* (Stockholm: Almqvist and Wiksell International, 1980), pp. 259–69.

24. Frank W. Notestein, "Demography in the United States: A Partial Account of the Development of the Field," *Population and Development Review*, vol. 8 (December 1982), pp. 651–58.

25. The complexities of Bush's views are evident in the committee report, largely written by Bush, advising James Bryant Conant, Harvard's president, on the disposition of the Gordon McKay bequest. See "Report of the Panel on the McKay Bequest to the President and Fellows of Harvard College" (Harvard University Printing Office, 1950).

26. Bush, *Science*, p. 19.

27. See Robert P. Multhauf, "The Scientist and the 'Improver' of Technology," *Technology and Culture*, vol. 1 (Winter 1959), pp. 38–47; and Thomas S. Kuhn, *The Essential Tension: Selected Studies in Scientific Tradition and Change* (University of Chicago Press, 1977), pp. 141–47.

28. Ralph E. Gomory and Roland W. Schmitt, "Science and Product," *Science*, vol. 240 (May 27, 1988), p. 1132, 1203. This point is also noted by Christopher Freeman, "Japan: A New National System of Innovation?" in *Technical Change and Economy Theory*, edited by Giovanni Dosi and others (Pinter, 1988), p. 346.

29. Multhauf, "The Scientist and the 'Improver' of Technology," p. 42.

30. Kuhn, *The Essential Tension*, p. 144.

31. Leonard S. Reich, *The Making of American Industrial Research: Science and Business at GE and Bell, 1876–1926* (Cambridge University Press, 1985), pp. 124–26.

32. See Bruno Latour, translated by Alan Sheridan and John Law, *The Pasteurization of France* (Harvard University Press, 1988). See also Claire Salomon-Bayet and others, *Pasteur et la Révolution Pastorienne* (Paris: Payot, 1986).

33. Judith P. Swazey and Karen Reeds, *Today's Medicine, Tomorrow's Science: Essays on Paths of Discovery in the Biomedical Sciences* (Department of Health, Education, and Welfare, 1978), especially chap. 4, pp. 53–72.

34. Bush, *Science*, p. 19.

35. Gomory and Schmitt, "Science and Product," pp. 1131–32, 1203. This point is also noted by Freeman, "Japan: A New National System of Innovation?" in *Technical Change and Economic Theory*, edited by Dosi and others, p. 346.

36. Rosenberg offers this charming anecdote from the career as a fundamental scientist that Edison foreswore: "In 1883 Edison observed the flow of electricity across a gap, inside a vacuum, from a hot filament to a metal wire. Since he saw no practical application, he merely described the phenomenon in his notebook and went on to other matters of greater potential utility in his effort to enhance the performance of the electric light bulb.

Edison was of course observing a flow of electrons, and the observation has since come to be referred to as the Edison effect. Had he been a patient scientist and less preoccupied with matters of short-run utility, he might later on have shared a Nobel Prize with Owen Richardson, who analyzed the behaviour of electrons when heated in a vacuum, or conceivably even with J J Thompson for the initial discovery of the electron." Nathan Rosenberg, "Critical Issues in Science Policy Research," *Science and Public Policy*, vol. 18 (December 1991), p. 337.

37. Bruce J. Hunt, "'Practice vs. Theory': The British Electrical Debate, 1888–1891," *Isis*, vol. 74 (September 1983), pp. 341–55.

Chapter Two

1. Benjamin Farrington, *Greek Science: Its Meaning for Us* (Penguin Books, 1953).

2. This dual position is admirably set forth in book VII by the exchange between Socrates and the hapless Glaucon, who abjectly fails to produce the appropriate reason for studying astronomy on his first two tries. Socrates has ultimately to explain that astronomical observations will be valuable only if they draw us away from the practical and empirical toward the invisible realities of the general forms. *The Republic of Plato: An Ideal Commonwealth*, rev. ed., translated by Benjamin Jowett (Willey Book Co., 1901).

3. G. E. R. Lloyd, *Early Greek Science: Thales to Aristotle* (W. W. Norton, 1970), p. 131.

4. *Metaphysica*, Book A, 1, p. 981b, lines 16–19, in W. D. Ross, ed., *The Works of Aristotle*, vol. 8, 2d ed. (Oxford: Clarendon Press, 1928).

5. Lloyd, *Early Greek Science*, p. 132.

6. *Metaphysica*, Book A, 2, lines 19–22.

7. G. E. R. Lloyd, *Greek Science after Aristotle* (W. W. Norton, 1973), p. 95.

8. A. C. Crombie, *Medieval and Early Modern Science*, vol. 1: *Science in the Middle Ages: v-xiii Centuries* (Doubleday Anchor Books, 1959), p. 5.

9. "Of the Parts of Animals," book I, chap. 1, *Historia Animalium*, in J. A. Smith and W. D. Ross, eds., *The Works of Aristotle*, vol. 4 (Oxford: Clarendon Press, 1910), pp. 468A–89A.

10. Crombie, *Medieval and Early Modern Science*, vol. 1, p. 65.

11. Edward Grant, *Physical Science in the Middle Ages* (John Wiley and Sons, 1971), p. 18.

12. C. P. Snow, *The Two Cultures: And a Second Look: An Expanded Version of the Two Cultures and the Scientific Revolution*, 2d ed. (Cambridge University Press, 1964), p. 32.

13. Charles Coulston Gillispie, *The Edge of Objectivity: An Essay in the History of Scientific Ideas* (Princeton University Press, 1990), p. ix.

14. Robert P. Multhauf, "The Scientist and the 'Improver' of Technology," *Technology and Culture*, vol. 1 (Winter 1959), p. 41.

15. Francis Bacon, *Novum Organum* (1620), book 1, aphorism 3, edited by Joseph Devey (P. F. Collier, 1901).

16. The quotation is from the Royal Society's second charter, issued by Charles II, April 22, 1663, under which it has ever since been governed. *Record of the Royal Society of London for the Promotion of Natural Knowledge*, 4th ed. (Edinburgh: Morrison and Gibb Ltd., 1940), p. 251.

17. For the interaction of science and technology over recent centuries, see Thomas S. Kuhn, *The Essential Tension* (University of Chicago Press, 1977), pp. 141–47. See also Multhauf, "The Scientist and the 'Improver' of Technology," pp. 38–47.

18. We ought not to overstate the purity of the natural philosphers' quest of understanding in these centuries. An interesting account of the relationship between Newton's theoretical mechanics and the technical needs of his time—in navigation, industry, and military affairs—is offered by B. Hessen, a Soviet scholar attending the 1931 London International Congress of the History of Science and Technology. However Marxist-Leninist his reasons for discovering a practical motive behind *Principia Mathematica*'s philosophic cover, Hessen gives a very clear-sighted analysis of the utility of Newton's mechanics for the technical problems of his day: "despite the abstract mathematical character of exposition adopted in the 'Principia' Newton was not only not a learned scholastic divorced from life, but in the full sense of the word was in the centre of the physical and technical problems and interests of his time." B. Hessen, "The Social and Economic Roots of Newton's 'Principia,'" in *Science at the Cross Roads: Papers Presented to the International Congress of the History of Science and Technology*, 2d ed. (London: Frank Cass and Company, 1971), p. 171. Yet the ideal of pure inquiry led Newton at the very least to cast his work in a philosophic style as devoid of direct practical content as the philosophic writings left behind by Archimedes were devoid of references to the engineering problems he solved for King Hiero.

19. H. Stafford Hatfield, *The Inventor and His World* (Penguin Books,1933), pp. 11–12.

20. Sir Eric Ashby, *Technology and the Academics: An Essay on Universities and the Scientific Revolution* (London: Macmillan, 1958), p. 50. See also A. E. Musson and Eric Robinson, *Science and Technology in the Industrial Revolution* (University of Toronto Press, 1969).

21. An irony of the industrial revolution in Britain is that a number of the practical men whose inventions brought them great wealth sought to gain for their sons access to the upper classes by sending them to the "public" schools, where they absorbed the Greek ideal of pure inquiry from the classical curriculum, as well as the anti-industrial values that dominated these classes. In *Marshalls of Leeds* (Cambridge University Press, 1960), William G. Rimmer details the history of an industrial firm that faltered because the second and third generations were lured away by the life of a country gentleman. For a more general analysis see "The Gentrification of the Industrialist," in Martin J. Wiener, *English Culture and the Decline of the Industrial Spirit 1850–1980* (Cambridge University Press, 1981), pp. 127–54.

22. For a fascinating account of the role of the needs of empire in Lord Kelvin's science, see Crosbie Smith and M. Norton Wise, *Energy and Empire: A Biographical Study of Lord Kelvin* (Cambridge University Press, 1989).

23. See Bruce J. Hunt, " 'Practice versus Theory': The British Electrical Debate, 1888–1891," *Isis*, vol. 74 (September 1983), pp. 341–55.

24. See Joseph Ben-David, "The Universities and the Growth of Science in Germany and the United States," *Minerva*, vol. 7 (Autumn–Winter 1968–69), pp. 1–35, and John S. Brubacher and Willis Rudy, *Higher Education in Transition: A History of American Colleges and Universities, 1636–1976*, 3d ed. (Harper and Row, 1976).

25. Multhauf, "Scientists and the 'Improvers' of Technology," *Technology and Culture*, p. 43. Multhauf is speaking here of more than the German experience alone.

26. W. K. Lewis notes the irony that although "the chemists turned out by the German universities were superlatively trained for these tasks . . . the very extent of the German success deflected professional development in the wrong direction." Dyestuffs were made in small-scale batch operations by chemists who continued to think in laboratory terms, and their somewhat outsize laboratory equipment was constructed by mechanical engineers with little understanding of the underlying chemistry. "The result was a divorce of chemical and engineering personnnel, not only in German technical industry but also in the universities and engineering schools that supplied that industry with professionally trained men." This, in Lewis's view, created an opening for the mass-production, continuous-flow methods of the chemical engineers in the United States who possessed the knowledge of chemistry, physics, and mechanics that was needed to control large-scale chemical reactions. See W. K. Lewis, "Chemical Engineering—A New Science," in Lenox R. Lohr, ed., *Centennial of Engineering: History and Proceedings of Symposia 1852–1952* (Chicago: Centennial of Engineering, Inc., 1953), pp. 696–99.

27. Cited by Ashby, *Technology and the Academics*, p. 25.

28. See Ben-David, "The Universities and the Growth of Science" pp. 1–35.'

29. Quoted by Multhauf in "The Scientist and the 'Improver' of Technology," p. 43.

30. Michael James Lacey, "The Mysteries of Earth-Making Dissolve: A Study of Washington's Intellectual Community and the Origins of American Environmentalism in the Late Nineteenth Century," Ph.D. dissertation, George Washington University, 1979.

31. Thomas Coulson, *Joseph Henry: His Life and Work* (Princeton University Press, 1950), pp. 69–72.

32. Charles C. Gillispie, *The Professionalization of Science*, The Neesima Lectures (Kyoto: Doshisha University Press, 1983), p. 36; and Charles C. Gillispie, *Science and Polity in France at the End of the Old Regime* (Princeton University Press, 1980), pp. 84–85.

33. George H. Daniels, "The Process of Professionalization in American Science: The Emergent Period, 1820–1860," in *Science in America since 1820*, edited by Nathan Reingold (New York: Science History Publications, 1976), pp. 63–78.

34. The significance of moving to more than one level of degree was partially lost on the American innovators who had gone to the German universities after completing their undergraduate studies in the United States. The educational experience they offered students admitted to the new American graduate schools was in this respect similar to their own. But it was very different from the experience of German students who studied for a single degree in the German universities.

35. Edwin Layton, "Mirror-Image Twins: The Communities of Science and Technology," in George H. Daniels, ed., *Nineteenth-Century American Science: A Reappraisal* (Northwestern University Press, 1972), pp. 210–30.

36. Multhauf, "The Scientist and the 'Improver' of Technology," p. 44.

37. Kuhn, *The Essential Tension*, p. 143. Emphasis added.

38. See the discussion of "the contrasting pattern of agricultural research" in Bruce L. R. Smith, *American Science Policy since World War II* (Brookings, 1990), p. 22.

39. A. Hunter Dupree, "Paths to the Sixties," in David L. Arm, ed., *Science in the Sixties* (University of New Mexico Press, 1965), p. 5.

40. Karl T. Compton, president of the Massachusetts Institute of Technology, James Bryant Conant, president of Harvard, Frank B. Jewett, president of Bell Labs and of the National Academy of Sciences, and Richard C. Tolman, dean of the graduate school of the California Institute of Technology.

41. See Daniel S. Greenberg, *The Politics of Pure Science* (New York: American Library, 1967).

42. Vannevar Bush disarmingly wrote in his autobiograpy *Pieces of the Action* (William Morrow and Co., 1970), pp. 31–32: "There were those who protested that the action of setting up N.D.R.C. was an end run, a grab by which a small company of scientists and engineers, acting outside established channels, got hold of the authority and money for the program of developing new weapons. That,

in fact, is exactly what it was." But this outside-channels initiative could not have succeeded if it had not matched Roosevelt's own fondness for creating new government agencies to get things done.

43. See George H. Daniels, "Office of Scientific Research and Development," in Donald R. Whitnah, ed., *The Greenwood Encyclopedia of American Institutions: Government Agencies* (Greenwood Press, 1983), pp. 426–32.

44. An excellent review of these, drawing extensively on documentary sources now available, appears in Daniel J. Kevles, "The National Science Foundation and the Debate over Postwar Research Policy, 1942–1945: A Political Interpretation of *Science—The Endless Frontier*," *Isis*, vol. 68 (March 1977), pp. 5–26.

45. See Greenberg, *The Politics of Pure Science*, pp. 51–67.

46. Few things so clearly attest to the growing authority of the scientific establishment than the contrast in the attention paid to private inventors in the two wars. One of the major efforts to enlist science and technology in World War I was the creation of a naval board, chaired by Thomas Edison himself, to review proposals from inventors. Harry Hopkins broached with Vannevar Bush the possibility of reviving such a board at the edge of World War II. Bush effectively killed it. See Bush, *Pieces of the Action*, p. 36.

47. How might the scientific information developed during the war be made known, consistent with military security, to advance the general welfare? What might be done to organize a program of medical research? What might the government do to aid research generally in public and private institutions? What might be done to discover and develop scientific talent in American youth? See Roosevelt's letter to Bush in Vannevar Bush, *Science—the Endless Frontier: A Report to the President on a Program for Postwar Scientific Research* (Washington: National Science Foundation, 1945, reprinted 1960), pp. 3–4.

48. The Bowman Committee's report, appendix 3 of the original Bush report, is reprinted in the edition of the Bush report that was published by the National Science Foundation on the occasion of the fortieth anniversary of NSF's founding. See Vannevar Bush, *Science—The Endless Frontier: A Report to the President on a Program for Postwar Scientific Research* (Washington: National Science Foundation, reprinted 1990), pp. 70–134. The initial capitalization of the National Research Foundation was to be $500 million. Interestingly, this Bowman Committee recommendation was echoed by the report of a National Academy of Sciences Committee, chaired by Harold Brown, in the late 1980s. The Brown Committee proposed that a non-profit R&D organization be created and provisioned by Congress with an up-front appropriation of $5 billion, which it would have full authority to spend over a period of years.

49. Kevles, "The National Science Foundation and the Debate over Postwar Research Policy, 1942–1945," p. 26.

50. National Science Foundation Act of 1950, P. L. 507, chapter 171, sec. 15(a), Security Provisions.

51. See *Biomedical Science and Its Administration: A Study of The National Institutes of Health* (Washington, 1965).

52. The two reports agreed on the importance of federal support for science and the development of a coherent science policy. But the Steelman Report laid

greater stress on the government's own scientific installations, which Bush had scarcely mentioned, and the need for the new scientific agency to be responsible to the president. In this vein, Steelman also recommended the appointment of a presidential science adviser. See President's Scientific Research Board, John R. Steelman, chairman, *Science and Public Policy: A Report to the President*, 5 vols. (GPO, 1947).

53. See "The Scientists," *Fortune*, vol. 38 (October 1948), p. 166. This statistic is cited in Kenneth Macdonald Jones, "Science, Scientists, and Americans: Images of Science and the Formation of Federal Science Policy, 1945–1950," Ph. D. dissertation, Cornell University, 1975, p. 362.

54. The importance of the cold war in sustaining the federal government's commitment to basic science is emphasized by Harvey Brooks, "National Science Policy and Technological Innovation," in Ralph Landau and Nathan Rosenberg, eds., *The Positive Sum Strategy: Harnessing Technology for Economic Growth* (Washington: National Academy Press, 1986), pp. 119–67.

55. See William A. Blanpied, ed., *Impacts of the Early Cold War on the Formulation of U.S. Science Policy: Selected Memoranda of William T. Golden, October 1950–April 1951* (Washington: American Association for the Advancement of Science, 1995).

56. For discussions of this point see Smith, *American Science Policy since World War II*, pp. 73–107; and Brooks, "National Science Policy and Technological Innovation," in Landau and Rosenberg, *The Positive Sum Strategy: Harnessing Technology for Economic Growth*, pp. 119–67.

57. C. W. Sherwin and R. S. Isenson, *First Interim Report on Project Hindsight (Summary)* (Department of Defense, Office of the Director of Defense Research and Engineering, 1966; final report, 1969). See also their "Project Hindsight: A Defense Department Study of the Utility of Research," *Science*, vol. 156 (June 23, 1967), pp. 1571–77.

58. See Illinois Institute of Technology Research Institute, *Technology in Retrospect and Critical Events in Science* [TRACES] (Washington: National Science Foundation, 1968). An interesting third perspective was offered by Gibbons and Johnston, who broadened this range of innovations by studying the role of science and scientists in a sample of recent product innovations by British firms selected from the "New Products" section of a current issue of each of the (complete set of) British technical journals. See Michael Gibbons and Ron Johnston, "The Roles of Science in Technological Innovation," *Research Policy*, vol. 3 (November 1974) pp. 220–42. Gibbons and Johnston characterize their findings as "midway between extremes set by Project Hindsight and TRACES," noting how often scientists were consulted by the problem solvers engaged in product innnovation. Their face-to-face interviews with these problem solvers made no effort to trace the up-stream, indirect influence of science on technical innovation. Such a backward mapping was the central feature of the most definitive American postscript to this debate, the Comroe and Dripps study, although this later work was limited to innovation in a particular branch of health care. See Julius H. Comroe Jr. and Robert D. Dripps, "Scientific Basis for the Support

of Biomedical Science," *Science*, vol. 192 (April 9, 1976), pp. 105–11. We return to the Comroe-Dripps findings in chapter 3.

Chapter Three

1. *Realising Our Potential: A Strategy for Science, Engineering and Technology*, Cm 2250 (London: Her Majesty's Stationery Office, 1993), p. 2.

2. National Science Foundation, *First Annual Report, 1950–51* (Washington: Government Printing Office, 1951), p. viii. Emphasis added.

3. Bush did not *define* basic research as work that is undertaken without thought of practical ends. Only if Bush had meant this description to be true in fact and not by definition would his famous dictum that applied research drives out pure be other than tautologically true.

4. Gerald Holton, *Science and Anti-Science* (Harvard University Press, 1993), p. 115.

5. Lillian Hoddeson, "The Emergence of Basic Research in the Bell Telephone System, 1875–1915," *Technology and Culture*, vol. 22 (July 1981), p. 514.

6. Deborah Shapley and Rustum Roy, *Lost at the Frontier* (ISI Press, 1985), p. 9

7. Julius H. Comroe Jr. and Robert D. Dripps, "Scientific Basis for the Support of Biomedical Science," *Science*, vol. 192 (April 9, 1976), pp. 105–11.

8. Alan T. Waterman, "The Changing Environment of Science," *Science*, vol. 147 (January 1, 1965), p. 15.

9. The full report appeared under the title *Applied Science and Technological Progress: A Report to the Committee on Science and Astronautics U.S. House of Representatives by the National Academy of Sciences* (Government Printing Office, 1967). Brooks's introductory chapter appeared in *Science* magazine in condensed form in advance of the report's appearance. See Harvey Brooks, "Applied Science and Technological Progress," *Science*, vol. 156 (June 30, 1967), pp. 1706–12. See also *Fundamental Research and the Policies of Government* (Paris: Organization for Economic Cooperation and Development, 1966), the report of an international study group, of which Brooks was a member, prepared for a meeting of the ministers of science of the OECD countries.

10. Brooks, "Applied Science and Technological Progress," p. 1706.

11. Ibid.

12. Ibid.

13. Ibid.

14. Ibid.

15. Britain too emerged from World War II deeply impressed by the contribution of its scientists to winning the war. Its gratitude took such tangible form as the government's swift approval of the 100-inch telescope proposed by the Royal Society to mark the tercentenary of Newton's birth.

16. OECD Directorate for Scientific Affairs, *The Measurement of Scientific and Technical Activities: Proposed Standard Practice for Surveys of Research and Experimental Development: Frascati Manual* (Paris: Organization for Economic

Cooperation and Development, 1962), p. 12. (Hereafter *Frascati Manual* and year of publication by OECD).

17. *Frascati Manual*, 1970, p. 13.

18. Ibid., pp. 13–15.

19. Ibid., p. 15.

20. *Frascati Manual*, 1981, p. 13.

21. John Irvine and Ben R. Martin, *Foresight in Science: Picking the Winners* (London: Frances Pinter Publishers, 1984); and Ben R. Martin and John Irvine, *Research Foresight: Priority-Setting in Science* (London: Pinter Publishers, 1989).

22. Martin and Irvine, *Research Foresight: Priority-Setting in Science*, p. 7.

23. National Science Foundation, *Report of the Task Force on Research and Development Taxonomy*, revised ed. (Washington, July 1989), p. 4.

24. Ibid., p. 3.

25. Ibid.

26. National Science Foundation, *Research and Development in Industry: 1989*, NSF 92–307 (Washington, 1990), p. 145.

27. *Frascati Manual*, 1993, p. 69.

28. Ibid.

29. Charles E. Falk, "Evaluation of Current Classifications of Research: A Proposal for a New Policy-Oriented Taxonomy," in Oliver D. Hensley, *The Classification of Research* (Texas Tech University Press, 1988), p. 153.

30. I first sketched such a framework in "Making Sense of the Basic/Applied Distinction: Lessons for Public Policy Programs," in *Categories of Scientific Research* (Washington: National Science Foundation, 1979), pp. 24–27, after setting it out for a council advisory to the director of the National Science Foundation. I argued the value of this framework in "Perceptions of the Nature of Basic and Applied Science in the United States," in Arthur Gerstenfeld, ed., *Science Policy Perspectives: USA-Japan* (Academic Press, 1982), pp. 1–18. I use "cell" and "quadrant" interchangeably here, although quadrant is, strictly speaking, less appropriate in view of the fact that the joint origin of this two-dimensional space lies at the lower left-hand corner rather than at the center of the table.

31. I. Arnon, *The Planning and Programming of Agricultural Research* (Rome: Food and Agriculture Organization of the United Nations, 1975), p. 29. A number of authors have commented on the willingness of firms to invest in basic research as a means of recruiting or upgrading staff, or of buying their way into scientific communications networks, or for other reasons apart from the knowledge the research may yield. See in particular, Nathan Rosenberg, "Why Do Firms Do Basic Research (With Their Own Money)?" *Research Policy*, vol. 19, no. 2 (April 1990), pp. 165–74.

32. Assembly of Behavioral and Social Sciences Study Project on Social Research and Development, *The Federal Investment in Knowledge of Social Problems* (Washington: National Academy of Sciences, 1978), pp. 55–56.

33. Comroe and Dripps, "Scientific Basis," pp. 105–11. See also their "Ben Franklin and Open Heart Surgery," *Circulation Research*, vol. 35 (November 1975), pp. 661–69.

34. Comroe and Dripps, "Scientific Basis," p. 106. See also p. 108.

35. Clearly the uses that partially inspired the research being assigned to Pasteur's quadrant here are by no means limited to the applications to cardiovascular or pulmonary disease from which Comroe and Dripps begin. Their study is in a sense a means of identifying a significant body of research which shows that the goals of understanding and use are substantially mingled.

36. Charles V. Kidd, "Basic Research—Description versus Definition," *Science*, vol. 129 (February 13, 1959), p. 368. Kidd cites the National Science Foundation study, *Basic Research: A National Resource* (Washington, 1957), p. 25.

37. John Irvine and Ben Martin, in their study of research foresight in the United States, found that the mission agencies of the federal government were more likely to report their support of research within utilitarian categories than were the academic recipients of this support. Thus the Department of Energy might describe as "engineering" research that the universities described as "physics."

38. Australian Science and Technology Council, *Basic Research and National Objectives* (Canberra: Australian Government Publishing Service, 1981), pp. 3–4. See also Australian Department of Science and Technology, *Project SCORE: Research and Development in Australia 1976–77* (Canberra: Australian Government Publishing Service, 1980), p. 457, and J. Ronayne, *Science in Government: A Review of the Principles and Practice of Science Policy* (Edward Arnold, 1984), p. 35.

39. I have borrowed this welcome phrase from Michael Heylin, "Science for the 21st Century," *Chemical and Engineering News* (March 13, 1995), p. 5.

40. Nathan Rosenberg, "Critical Issues in Science Policy Research," *Science and Public Policy*, vol. 18 (December 1991), p. 335.

41. Ryo Hirasawa, "The Concept of R&D Management Sprouting in Japan," in *Proceedings of the Research and Technology Planning Society* (May 15, 1992).

42. Hirotaka Takeuchi and Ikujiro Nonaka, "The New New Product Development Game," *Harvard Business Review*, vol. 64 (January–February, 1986), pp. 137–46. As further evidence of the importance of metaphorical thinking for these management developments, Takeuchi and Nonaka note that Fuji-Xerox identifies its departure from the linear, sequential mode of development as the sashimi system, sashimi being slices of raw fish arranged in an overlapping pattern on a dish.

43. Stephen J. Kline, "Innovation Is Not a Linear Process," *Research Management*, vol. 28 (July–August 1985), pp. 36–45. See also Stephen J. Kline and Nathan Rosenberg, "An Overview of Innovation," in Ralph Landau and Nathan Rosenberg, eds., *The Positive Sum Strategy: Harnessing Technology for Economic Growth* (Washington: National Academy Press, 1986), pp. 275–305.

44. Erich Bloch and David Cheney, "Technology Policy Comes of Age," *Issues in Science and Technology*, vol. 9 (Summer 1993), p. 57.

45. *An Integrated Approach to European Innovation and Technology Diffusion Policy: A Maastricht Memorandum*, May 1993.

46. Nathan Rosenberg, "The Impact of Technological Innovation: A Historical Review," in Landau and Rosenberg, *The Positive Sum Strategy*, pp. 17–32.

47. The meaning of "trajectory" is closely similar to its meaning in the work of Richard R. Nelson and Sidney G. Winter. See, for example, "In Search of Useful Theory of Innovation," *Research Policy*, vol. 6 (January 1977), pp. 36–76, and chap. 11 of their work, *An Evolutionary Theory of Ecomomic Change* (Harvard University Press, 1982).

48. See Harvey Brooks, "The Relationship between Science and Technology," *Research Policy*, vol. 23 (September 1994), p. 479.

Chapter Four

1. Daniel J. Kevles quotes Don K. Price as saying that Senator Harley Kilgore's more populist notions of science "are slipping up on us again" as the country's mood changed in the 1960s. See Daniel J. Kevles, "The Crisis of Contemporary Science: The Changed Partnership," *Wilson Quarterly*, vol. 19 (Summer 1995), p. 47. Excellent reviews of the broad stages of postwar science policy are provided by Bruce L. R. Smith, *American Science Policy since World War II* (Brookings, 1990) and Harvey Brooks, "National Science Policy and Technological Innovation," in Ralph Landau and Nathan Rosenberg, eds., *The Postive Sum Strategy: Harnessing Technology for Economic Growth* (Washington: National Academy Press, 1986), pp. 119–67.

2. *Report of the Task Force on the Health of Research*, chairman's report to the House Committee on Science, Space, and Technology, 102 Cong. 2 sess., serial L (Government Printing Office, 1992), p. 2.

3. *Realising Our Potential: A Strategy for Science, Engineering and Technology*, Cm 2250 (London: Her Majesty's Stationery Office, 1993).

4. Departments of Veterans Affairs and Housing and Urban Development and Independent Agencies Appropriation Bill, 1994, S. Rept. 137, 103 Cong. 1 sess. (Government Printing Office, 1993), p. 4.

5. These figures are taken from Paul Krugman, "Growing World Trade: Causes and Consequences," unpublished paper, March 1995, draft prepared for the Brookings Panel on Economic Activity, April 1995.

6. Vannevar Bush, *Science—the Endless Frontier: A Report to the President on a Program for Postwar Scientific Research* (Washington: National Science Foundation, reprinted 1990), p. 19.

7. H. Con. Res. 67, *Congressional Record*, June 29, 1995, pp. H6561–6583.

8. See especially Jon D. Miller, *The American People and Science Policy: The Role of Public Attitudes in the Policy Process* (Pergamon Press, 1983); Jon D. Miller and Kenneth Prewitt, *A National Survey of the Non-governmental Leadership of American Science and Technology: A Report to the National Science Foundation*, unpublished report of the Public Opinion Laboratory of Northern Illinois University, May 1982; Kenneth Prewitt, "The Public and Science Policy," *Science, Technology, and Human Values*, vol. 7 (Spring 1982), pp. 5–14; and Jon

D. Miller and Linda K. Pifer, *The Public Understanding of Biomedical Science in the United States, 1993* (Chicago: Chicago Academy of Sciences, 1995). For the most recent indicators of public opinion biennially published by the National Science Board, see "Science and Technology: Public Attitudes and Public Understanding," in National Science Board, *Science and Engineering Indicators* (GPO).

9. Miller and his colleagues use a variety of measures to stratify the public according to the layers of this pyramid. At the apex, *science policy decisionmakers* are those "with power to make binding decisions" on science policy. *Nongovernment science policy leaders* are those with "a strong organizational base within the scientific community." Each of these groups is a tiny element of the country as a whole. *The attentive public* (less than 20 percent of the public) has a high level of interest in science matters as well as a functional level of knowledge about science. *The interested public* (another 20 percent) are those with high interest who do not have such a functional level of knowledge. *The nonattentive public* (60 percent of the public) constitutes the remainder of the adult public. See Miller, *The American People and Science Policy*, pp. 33–49.

10. Miller, *The American People and Science Policy*, p. 52. The comparable figures from a 1979 survey were 88 percent for the attentive public, 74 percent for the interested public, and 63 percent for the nonattentive public.

11. Miller, *The American People and Science Policy*, p. 53.

12. See the Harris Poll for February–March of 1994 and the National Science Board's *Science and Engineering Indicators–1993*, drawing on J. D. Miller and L. K. Pifer's *Public Attitudes toward Science and Technology, 1979–1992, Integrated Codebook* (Chicago: International Center for the Advancement of Scientific Literacy, Chicago Academy of Sciences, 1993) and unpublished tabulations. These findings are the more impressive in view of the distant and, in some respects, unfavorable aspects of public perceptions of science. For example, the Harris Poll sample divided 51 percent to 47 percent on the statement, "I understand less and less of what scientists are doing today," and the surveys reported by the National Science Board found 52 percent of all adults agreeing that "many scientists make up or falsify research results to advance their careers or make money."

13. National Academy of Sciences, *Science, Technology, and the Federal Government: National Goals for a New Era* (Washington: National Academy Press, 1993).

14. Ralph E. Gomory, "An Unpredictability Principle for Basic Research," The 1995 William D. Carey Lecture in American Association for the Advancement of Science, *Science and Technology Yearbook 1995* (Washington, 1995), pp. 3–17.

15. Insight into this relationship is provided by Henry Ehrenreich, "Strategic Curiosity: Semiconductor Physics in the 1950s," *Physics Today*, vol. 48 (January 1995), pp. 28–34.

16. Maurice B. Strauss, *Familiar Medical Quotations* (Little, Brown, 1968), p. 519.

17. Report of the Panel on the McKay Bequest to the President and Fellows of Harvard College (1950), p. 7.

18. Harvey Brooks, "The Relationship between Science and Technology, *Research Policy*, vol. 23 (September 1994), p. 482.

19. This is not to suggest that either Staudinger or Carothers was narrowly targeted on industrial uses. Each was a distinguished Pasteur's quadrant scientist. Carothers led the development at Dupont of a laboratory dedicated to fundamental research "without any regard or reference to commercial objectives," while Staudinger was for many years preoccupied by the often bitter struggle to establish that polymers were in fact huge macromolecules held together by the same forces as were found within their repeating monomers rather than by the colloidal forces posited by association theory.

20. Harold T. Shapiro, "The Evolution of U.S. Science Policy," unpublished paper, August 1993.

21. It is hardly likely that so central an issue could drop entirely out of sight, and a number of writers have dealt with parts of it. See, for example, Alvin M. Weinberg, "Criteria for Scientific Choice," *Minerva*, vol. 1 (Winter 1963), pp. 159–71, and Harvey Brooks, "The Problem of Research Priorities," *Daedalus*, vol. 107 (Spring 1978), pp. 171–90.

22. By far the best comparative survey of national practice is the book-length study by Ben R. Martin and John Irvine, *Research Foresight: Priority-Setting in Science* (London and New York: Pinter, 1989). Their survey encompasses eight countries, including France, Germany, and the United States, as well as Japan.

23. See Government of Japan, *Basic Policy for Science and Technology*, April 24, 1992 (translation by the Science and Technology Policy Bureau of the Science and Technology Agency), pp. 5–6.

24. There were substantial barriers to increased investment in university research when the government resolved to increase its expenditures on basic science. Partly owing to the ideological leanings and resistance to use among the science faculties of universities formed on the German model, there was considerable estrangement between these institutions and the government during the long period of Liberal Democratic rule, and the universities remained an isolated province administered by the Ministry of Education, Science, and Culture. This ministry was late in developing a system of competitive grants to gifted university scientists and continued to distribute most of its research funds through the highly inefficient, seniority-based *koza* system.

25. This was true of Tonomura's use of electron holography to explore superconductivity at Hitachi's Advanced Research Laboratories in Hatoyama Prefecture and Pepper's effort to develop the scientific knowledge that would allow semiconductors to be built atomic layer by atomic layer at Toshiba's Industrial Research Laboratory in Cambridge, England, one of several facilities these global firms established outside Japan.

26. The April 1992 Cabinet decision closely coupled an endorsement of increased investment in basic science with a list of sixteen R&D priority areas judged promising in terms of both needs and seeds. The concern for potential use is also revealed in the interdisciplinary character of many of the areas marked off as priorities, such as "biotronics" and "chematronics," fields that link elements

of information technology with biology and with chemistry and materials science. See Government of Japan, *Basic Policy for Science and Technology.*

27. The lead in this effort has been taken by Hariolf Grupp of the Frauenhofer Institut in Karlsruhe. See Hariolf Grupp (in cooperation with Sibylle Breiner, Kerstin Cuhls, and Ben Martin), *Methodology for Identifying Emerging Generic Technologies—Recent Experiences from Germany, Japan and the USA* (Karlsruhe: Frauenhofer Institute for Systems and Innovations Research, 1992).

28. Nigel Williams, "U.K. Tries to Set Priorities with the Benefit of Foresight," *Science*, vol. 268 (May 12, 1995), pp. 795–96.

Chapter Five

1. It will also be useful to keep in view another vertical dimension, the distinction between little and big science. This further spectrum is partially captured by one of the typologies by which the federal government keeps its research accounts, the categories of "modes of support," which vary from support to individual investigators, to research teams of senior investigators who are often at different institutions, to research centers, and to major facilities that are of regional or national scope. Although these two vertical dimensions are by no means the same, they are closely aligned, since the support of research centers or of regional or national facilities can be a form of research wholesaling.

2. For an excellent review of this detective story and the reception of its findings into the policy process, see W. Henry Lambright, "NASA, Ozone, and Policy-Relevant Research," *Research Policy*, vol. 24 (September 1995), pp. 747–60.

3. The Rothschild Report, "The Organization and Management of Government R. and D.," was included in the Government's Green Paper, *A Framework for Government Research and Development*, Cmnd 4814 (London: Her Majesty's Stationery Office, 1971). The Government's White Paper is *Framework for Government Research and Development*, Cmnd 5046 (London: Her Majesty's Stationery Office, 1972).

4. See Harriett Zuckerman and Robert K. Merton, "Institutionalized Patterns of Evaluation in Science 1971," in Robert K. Merton, ed., *The Sociology of Science: Theoretical and Empirical Investigations* (University of Chicago Press, 1973), p. 463.

5. Testimony by Ruth L. Kirschstein, director of National Institute of General Medical Sciences, NIH, HHS, Hearings on Research Project Selection before the Task Force on Science Policy of the House Committee on Science and Technology, April 8–10, 1986, 99 Cong. 2 sess. (Government Printing Office, 1986), vol. 17, pp. 64–69, especially p. 65.

6. See in particular the book-length critique by Daryl E. Chubin and Edward J. Hackett, *Peerless Science: Peer Review and U.S. Science Policy* (Albany, N.Y.: State University of New York Press, 1990).

7. Especially comprehensive are these reports: NIH Grants Peer Review Study Team, *Grants Peer Review: Report to the Director, NIH Phase I* (Washington,

December 1976); Stephen Cole, Leonard Rubin, and Jonathan R. Cole, *Peer Review in the National Science Foundation: Phase One of a Study* (Washington: National Academy of Sciences, 1978); Jonathan R. Cole and Stephen Cole, *Peer Review in the National Science Foundation: Phase Two of a Study* (Washington: National Academy Press, 1981); National Science Foundation, *Final Report: NSF Advisory Committee on Merit Review 1986* (Washington, 1986); and National Science Foundation, *Proposal Review at NSF: Perceptions of Principal Investigators*, NSF Report 88-4 (Washington, 1988).

8. An example of this congressional inclination that capitalizes on the nature of representation in the U.S. Senate is the requirement that a specified fraction of research funds go to a group of states that are light in population and established research institutions, a requirement that has acquired the acronym EPSCOR.

9. "FY 1996 Research and Development Priorities," memorandum from John H. Gibbons and Leon E. Panetta for the heads of executive departments and agencies, May 6, 1994.

10. It should be noted that responsibility for research on weather modification was part of the original NSF charter. Indeed, Congress's interest in quick results and premature scale-up of operations probably reinforced the National Science Board's early reluctance to take responsibility for programs of mission-oriented research.

11. An excellent review of this development is provided by Dian Olson Belanger, *Enabling American Innovation: Engineering and the National Science Foundation* (Purdue University Press, forthcoming).

12. The National Science Foundation included this question in a November 1986 survey of principal investigators who had received NSF grants: "Would you say that all or much of the research carried out under the NSF award has 'applied,' 'practical' or 'policy' implications beyond the advancement of knowledge itself?" Foundation officials were startled to find that 53 percent of respondents answered yes. Of those who did, 24 percent went on to say that these implications were "clear and immediate," 53 percent that they were "fairly clear but long range," and 24 percent that they were "potential but not clear." See National Science Foundation, *Proposal Review at NSF: Perceptions of Principal Investigators*; and Jim McCullough, "First Comprehensive Survey of NSF Applicants Focuses on Their Concerns about Proposal Review," *Science, Technology, and Human Values*, vol. 14 (Winter 1989), pp. 78–88.

13. *A Foundation for the 21st Century: A Progressive Framework for the National Science Foundation,* Report of the National Science Board Commission on the Future of the National Science Foundation (Washington: National Science Foundation, 1993), p. 5.

14. *Departments of Veterans Affairs and Housing and Urban Development and Independent Agencies Appropriations Fiscal Year 1994,* Hearings before the Senate Committee on Appropriations, 103 Cong. 1 sess. (Government Printing Office, 1993), pt. 1, p. 12.

15. Departments of Veterans Affairs and Housing and Urban Development, and Independent Agencies Appropriation Bill, 1994, S. Rept. 137, 103 Cong. 1 sess. (GPO, 1993), p. 168. Senator Mikulski and others seeking to direct more

of the country's research strength to unmet societal needs soon learned to displace "strategic research" in favor of a term such as "fundamental research in areas of strategic national importance," a phrasing echoing Holton's *"research in an area of basic scientific ignorance that lies at the heart of a social problem."* Emphasis in original. See Gerald Holton, *Science and Anti-Science* (Harvard University Press, 1993), p. 115.

16. For much of this account I am indebted to Belanger, *Enabling American Innovation.*

17. Belanger, *Enabling American Innovation,* quoting NSB Chair Philip Handler.

18. The amendment to the Military Procurement Authorization Act of 1970 sponsored by the Senate's majority leader, Mike Mansfield, is a classic example of a limited action with far broader effects. His section 203 required that none of the funds authorized by the act "may be used to carry out any research project or study unless such project or study has a direct and apparent relationship to a specific military function or operation." It applied only to defense appropriations and only for a year; indeed, more permissive language was already incorporated in the Military Procurement Authorization Act of 1971. But the response to its clear message by the other mission agencies of the government led to OMB's proposal to NSF. And by 1974 the National Science Board was itself so fearful of the Foundation's being overwhelmed by basic research cast adrift by the other agencies that it set aside its traditional reluctance to assume responsibility for science policy across the government and urged the agencies to increase their support for basic research of even potential relevance to their missions. See Bruce Smith, *American Science Policy since World War II* (Brookings, 1990), pp. 81–82.

19. Belanger, *Enabling American Innovation.*

20. This section's figures for basic research in the mission agencies are taken from Intersociety Working Group, *AAAS Report XX: Research and Development FY 1996* (Washington: American Association for the Advancement of Science, 1995), p. 56, table 1-8.

21. See Stuart W. Leslie, *The Cold War and American Science: The Military-Industrial Academic Complex at MIT and Stanford* (Columbia University Press, 1993), pp. 188–211.

22. Evidence of how important this implicit bargain was in the thinking of the Atomic Energy Commission is furnished by William Golden's survey of science policy for the Truman administration. When Golden asked Kenneth Pitzer, director of AEC's research division, how much of AEC's basic research might be turned over to the newly created National Science Foundation, Pitzer indicated that the fraction would be small. But Golden's record of the conversation noted that Pitzer went on to say that "there would be a strong tendency to hold control over basic research activities in institutions which were also performing classified programmatic research for the AEC, for in general the unclassified basic research work is much more attractive to universities." From a memo for the file, November 1, 1950. See William A. Blanpied, *Impacts of the Early Cold War on the Formulation of U.S. Science Policy: Selected Memoranda of William T. Golden,*

October 1950–April 1951 (American Association for the Advancement of Science, 1995), p. 18.

23. National Research Council, *Plasma Science: From Fundamental Research to Technological Applications* (Washington: National Academy Press, 1995), p. 9.

24. Ibid., p. 22.

25. This pressure is described by the Office of Technology Assessment's report on the Tokamak Physics Experiment (TPX) and other approaches. See U.S. Congress, Office of Technology Assessment, *The Fusion Energy Program: The Role of TPX and Alternate Concepts,* OTA-BP-ETI-141 (Government Printing Office, 1995).

26. National Research Council, *Plasma Science,* p. 145. Emphasis in original.

27. For evidence that recent computer simulations have begun to build up a quantitative first-principles understanding of transport and turbulence in magnetically confined plasmas, see M. Kotschenreuther and others, "Quantitative Predictions of Tokamak Energy Confinement from First-Principles Simulations with Kinetic Effects," *Physics of Plasmas,* vol. 2 (June 1995), pp. 2381–89.

28. National Research Council, *Plasma Science,* p. 146.

29. In a single decade the fraction of GNP the United States spent on health care rose from 9.1 percent in 1980 to 11.6 percent in 1989. The margin between the annual compound growth rate of spending on health and on all nonhealth sectors of the economy is an astonishing 3.1 percentage points. See Victor R. Fuchs, "The Health Sector's Share of the Gross National Product," *Science,* vol. 247 (February 2, 1990), pp. 534–38.

30. John Sherman, "A History of the Politics of Health Research," paper presented at the Science and Technology Policy Colloquium, American Association for the Advancement of Science, 1992.

31. Kaplan remarks that "in addition to the social pressures from the scientific community there is also at work a very human trait of individual scientists. I call it the law of the instrument, and it may be formulated as follows: Give a small boy a hammer, and he will find that everything he encounters needs pounding. It comes as no particular surprise to discover that a scientist formulates problems in a way which requires for their solution just those techniques in which he himself is especially skilled." See Abraham Kaplan, *The Conduct of Inquiry: Methodology for Behavioral Science* (Chandler Publishing Company, 1964), p. 28.

32. An excellent review of the benefits of pluralism appears in Rodney W. Nichols, "Pluralism in Science and Technology: Arguments for Organizing Federal Support for R&D around Independent Missions," *Technology in Society,* vol. 8, nos. 1–2 (1986), pp. 33–63.

33. See Daniel J. Kevles, "The Crisis of Contemporary Science: The Changed Partnership," *Wilson Quarterly,* vol. 19 (Summer 1995), pp. 41–52, quotation on p. 48.

34. D. Allan Bromley, *The President's Scientists: Reminiscences of a White House Science Advisor* (Yale University Press, 1994), p. 142.

35. See in particular the prospectus, *A Proposal for a National Institute for the Environment: Need, Rationale, and Structure* (Washington: Committee for the National Institute for the Environment, September 1993). The Department of Energy has also supported a National Institute of Global Environment in which regional boards have put together packages of proposals, although the experience of NIGEC shows how easily distributive politics can overwhelm scientific criteria, as in the creation of its politically expedient Alabama-based southeastern center out of its New Orleans-based southern regional center.

36. President's Scientific Research Board, John R. Steelman, chairman, *Science and Public Policy: A Report to the President*, 5 vols. (GPO, 1947).

37. In his review of U.S. science for the Truman administration, William T. Golden was the first to advocate a permanent position of science adviser to the president in peacetime. See Blanpied, *Impacts of the Early Cold War on the Formulation of U.S. Science Policy.*

38. He described these in two articles published soon after leaving office. See Frank Press, "Science and Technology in the White House, 1977 to 1980: Part 1 and Part 2," *Science*, vol. 211 (January 9 and 16, 1981), pp. 139–45, and 249–56.

39. There is a mild anomaly in designating as one of the assistants to the president a White House official who also held a post requiring Senate confirmation and could therefore be required to testify on Capitol Hill. The dictum that assistants to the president should have a passion for anonymity has long since given way to the need of presidents for high-profile assistants to deal with cabinet secretaries, congressional leaders, and other senior figures for whom there are not enough minutes in a president's day. But there was unease that *this* assistant to the president might be asked by Congress to testify about confidential White House discussions. D. Allan Bromley gives an amusing account of how this problem was in his case solved by housing him in the Old Executive Office Building rather than in the west wing of the White House with the assistants to the president not appointed by and with the advice and consent of the Senate. See Bromley, *The President's Scientists*, pp. 44–45.

40. Although the Bush administration was sharply criticized by environmental activists and the press for allowing the United States to become isolated from most of the industrial world in the Rio conference, it took several steps to strengthen the scientific foundations of environmental policy. One of these clarified the economic aspects of environmental policy that could help guide the research on climate change coordinated by the Climate Change Working Group of the Domestic Policy Council, an effort that highlighted the importance of the input to models of global change from the satellite data to be generated by NASA's Mission to Planet Earth Program. Global change research also became one of the "cross-cuts" that were designated as formal presidential initiatives in the annual submissions to Congress developed by the corresponding interagency groups of the Federal Coordinating Council for Science, Engineering, and Technology.

41. Prominent among these priorities are high-performance computing and communication, advanced materials and processing, biotechnology, and the U.S.

Global Change Research Program, as well as an educational priority in mathematics and science.

42. This development had a notable assist from the Congressional Fellowship Program of the American Association for the Advancement of Science, which opened important new career vistas for a number of scientists or engineers with a strong interest in science and technology policy.

43. This and other points on which the Office of Technology Assessment was vulnerable are developed in a very perceptive requiem for the agency offered by Fred W. Weingarten, a former member of its staff. See Weingarten, "Obituary for an Agency," *Communications of the ACM*, vol. 38 (September 1995), pp. 29–32.

44. This lesson is highlighted by a former OTA official's account of having tried to sell a study of satellite communications to the senior aide of a key senator. The aide's response was that "the Senator has already made up his mind on that issue." "Wouldn't he like to know whether it was technologically feasible?" the OTA official persisted. "Of course not," the aide snapped. "Why would he?" See Weingarten, "Obituary for an Agency," p. 31.

45. Elmer B. Staats, "Reconciling the Science Advisory Role with Tensions Inherent in the Presidency," pp. 79–96, in William T. Golden, *Science Advice to the President,* a special issue of *Technology in Society* (Pergamon Press, 1980), pp. 87–88.

46. Cited by Staats in "Reconciling the Science Advisory Role with Tensions Inherent in the Presidency," p. 91, note 15.

INDEX

174